1985

SYSTEMS, MANAGEMENT AND CHANGE a graphic guide

SYSTEMS, MANAGEMENT AND CHANGE

RUTH CARTER, JOHN MARTIN,
BILL MAYBLIN AND MICHAEL MUNDAY

a graphic guide

Harper & Row Publishers
in association with
The Open University

Picture Credits

Page 1, from left to right
BBC Hulton Picture Library
Popperfoto
Black Star

Page 7
Reproduced by permission of Geographers' A-Z Map Co. Ltd.
Based upon the Ordnance Survey map with the permission of
the controller of Her Majesty's Stationery Office, Crown
copyright reserved.

Page 9
Reproduced by permission of British Telecom

Page 12
V A W Hillier *Motor Vehicle Basic Principles* (1976) Hutchinson

Page 17
Reproduced with the permission of the controller of Her
Majesty's Stationery Office, Crown Copyright reserved.

Page 33
V A W Hillier *Motor Vehicle Basic Principles* (1976) Hutchinson

Page 44
Ann Ronan Picture Library

Page 97
Popperfoto

All other photographs by Rogue Images

All illustrations by Bill Mayblin and Michael Munday

Harper & Row Ltd
28 Tavistock Street
London WC2E 7PN

British Library Cataloguing in Publication Data

Systems, management and change.
1. Sequential analysis
2. Decision-making-Mathematical models
I. Carter, Ruth
519,5'42   QA279.7
ISBN 0 06 318 272 6

Printed and bound by Hollen Street Press

Acknowledgements

This book could not possibly have come into existence except as a team product. Even amongst the four of us listed on the title page, the roles of manager, designer, writer and drawer merged and shifted continuously, even though there were specialists in each; and, of course, we had the detailed, careful and good humoured support of June McGowan who typed it all many times, and did battle with the word-processor! The design input to the book was made possible through the human and technical resources of Information Design Workshop, Rotherhithe, London.

But though the core team created the book, and have the direct responsibility for what appears in it, they have been profoundly influenced and challenged by many others. The book arose from the teaching needs of the Open University Systems Group, and in particular the team for Course T301 ('Complexity, management and change: applying a systems approach'); many members of the Systems Group provided reactions and comments; and we have drawn freely on teaching material and experience that is really the communal property of the Group.

We received invaluable help from our team of developmental testers, drawn mainly from students of the Open University's Yorkshire Region. Many friends and colleagues outside the Open University Systems Group also commented on drafts: Peter Checkland, Ranulph Glanville, Carol Halliwell, Colin Hampson-Evans, Nimal Jayaratna, Lyn Jones, John Robson and his students at Luton College, Chris Saxton, Barry Turner, John Watt, Ian Woodburn. To all of these, our thanks — and our apologies to anyone we have inadvertently omitted.

John Martin would also like to acknowledge his immense debt to his trainers and fellow trainees at the Boyeson Centre, and to David L. Smith. They will certainly recognise their influence.

Finally, we would like to acknowledge our debt to the books of Don Koberg and Jim Bagnall, of Richard Appiganesi and Oscar Zarate, and of Raymond Briggs.

Your comments

We should be very glad to hear readers' comments on the book, so that future editions can be improved. Send them to John Martin, Systems Group, Open University, Milton Keynes, Buckinghamshire, MK7 6AA, UK.

1.

hange..

. . . occurs continuously all around us. We may want and support it, fear it, be indifferent to it, be passive, or participate in it.

Activity is always messy, though there is often some order.

This book is about exerting influence on the messiness of change — at work, at home, in the community, in government . .

Change and the broad view.

There are different ways of thinking about change

— The biologist might see it in evolutionary terms.

— The sociologist might see the development of social conflict.

— The economist might see it in terms of market forces.

We are looking at it as the result of <u>responsible actions:</u> an agent is a person who tries to understand the situation, attempts to allow for the unknowns, acts accordingly, is aware of the results, and accepts responsibility for them.

    All practical decision-makers are agents, responsible for the <u>full</u> effects of their actions — not just the hoped-for ones. So 'responsible action' involves seeing the situation <u>as a whole.</u>

WE DO NEED SPECIALIST SKILLS:

    * The in-depth study of
                  one feature

    * The technical skills to
              get the details right

But, though a situation with, say, five or six independent components is something you might get to grips with, one with twenty five is beyond even intuitive judgement. Yet ultimately everything tends to be connected to everything else, so the world is an immensely complicated place.

It's no good expecting computers to overcome this, because though they can manipulate very large numbers of components, manipulation is not the same as understanding. Nor can you rely on big teams that bring many brains together, since the need to understand what the others are doing still limits the complexity of what the team can handle.

So 'seeing the situation as a whole' can't mean 'thinking of everything'.

BUT WE ALSO NEED
     ROUNDED UNDERSTANDING:

   * Looking at it from many
                        angles
   * Trying to see the wood
            in spite of the trees
   * Searching for the heart
           of the matter

The language of Systems

There are different ways of talking about 'wholes'.

We can talk about how wholeness feels: — the insightfulness, depth, richness or fidelity of a description, or the sense of a complete, aesthetically satisfying, shape or structure.

We can talk of wholeness as a pattern in a 'field' of forces rather like the pattern of iron filings in a field round a magnet, as when we talk of 'tensions' which can be 'resolved', or of competing 'forces' that 'draw us' along particular 'lines of action'. 'Wholeness' is then a matter of being aware of all the 'forces' in the situation, so that, like someone canoeing a fragile boat up powerful rapids, we can work with them to achieve our goal.

However this book will use a third 'language' for describing wholes, using the concept of a bounded system of linked components. A situation described in this 'language' is represented as:

a collection of elements that represent the relatively fixed parts of the situation, at the finest level of analysis that we want to go to. Here are some elements:

which we will represent as symbols:

(RUTH)   (RUTH'S PHONE)   (EXCHANGE)   (JOHN'S PHONE)   (JOHN)

which may be grouped into sub-systems to give a picture of the structure of the situation:

((RUTH)   (PHONE)   RUTH'S HOUSE)   (EXCHANGE)   ((PHONE)   (JOHN)   JOHN'S HOUSE)

we then add <u>connecting links,</u> that indicate the changing flows, influences and causal connections that bring the structure to life — its '<u>process</u>':

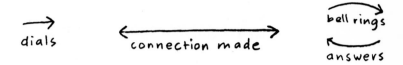

so as to produce a picture of an organised, active, assembly:

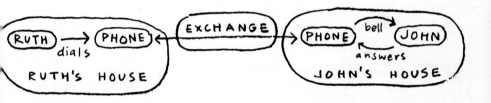

But there are lots of things in Ruth's house, John's house, and the world at large that have little or nothing to do with Ruth phoning John, so we need a <u>boundary</u> to mark off those <u>components</u> (general term for sub-systems or elements) that are to be treated as part of the system, and those that are not:

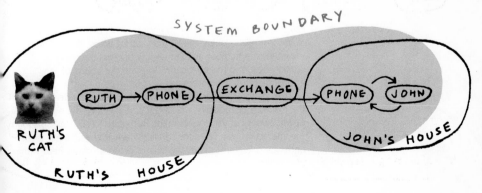

Four principles for drawing boundaries

1. Exclude components or relationships that have no functional effect on the system relevant to its descriptive purpose.

    If the presence or absence of Ruth's cat makes no difference to Ruth phoning John, then it is not part of the Ruth-phoning-John system.

2. Items that <u>can be strongly influenced or controlled</u> by the system or its owner should clearly be included in the system because you must understand how they work. Items that <u>influence</u> the system, but <u>cannot be influenced or controlled by it</u> may be better put outside the system, in its <u>environment</u>; you only need to know their effects, and excluding them helps to keep your detailed system analysis to a manageable size.

    High charge rates lead to shorter phone calls; clearly they influence Ruth-phoning-John. But Ruth and John do not need to know the mechanics of charging, and cannot influence the rate, so this factor is best relegated to the environment.

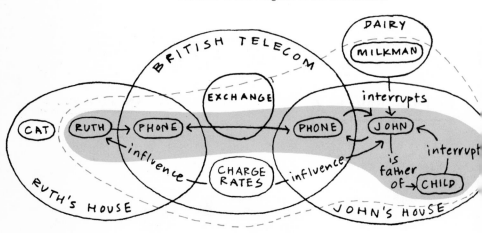

'RUTH-PHONING-JOHN' SYSTEM

--- LIMIT OF ENVIRONMENT

3. Position the boundary either to enclose or to exclude <u>complete</u> clusters of relationships, rather than cutting across them. This minimises the number and complexity of cross-boundary relationships, and makes it easier to grasp the effects of the environment on the system.

    Though 'the milkman interrupts John', a deeper understanding of the milkman or the dairy would not add anything; the milkman clearly belongs in the environment. But when 'child interrupts John' the interaction may be more significant; we may well need to understand its implications: the child is probably better kept within the system.

4. A useful description will usually depict the situation as neither totally 'open' nor totally 'closed', but somewhere in between, because:

—<u>a totally open</u> system would be one in which the environment is so important that the system merges into it, has an arbitrary boundary, has no stable identity, and is therefore very hard to manage or plan for.

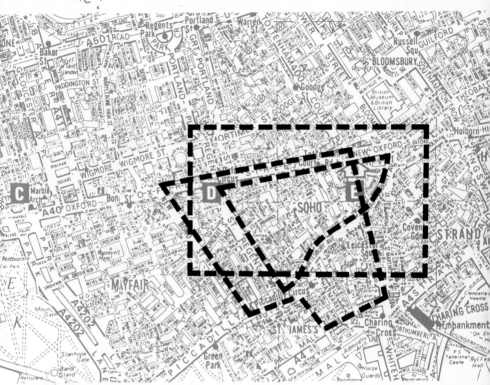

London's West End: where do you draw the line?

Even an egg is not a totally closed system

A <u>totally closed</u> system would be self-contained, with no environment at all. It could not be influenced by external events, you could not intervene in it, and it could not serve any useful external purpose.

However some useful systems may need to close <u>temporarily</u> (e.g. for dormant periods, defensive retreat, internal reorganisation, stock-taking etc.) or may have relatively closed <u>sub-systems</u> (e.g. frameworks that provide a fixed structure for the rest of the system).

7

Systems descriptions relate to particular view-points

When I try to describe a situation as a system, I am trying to find a way of thinking about it that will help me to see how it could achieve something for somebody — perhaps me, perhaps someone else.

Though I will want to give my system a convenient name, so that I can talk about it and think about it more easily, I am not naming a completely objective entity, like a football, or an elephant, which everyone would agree about.

A system's description is partly subjective and 'private':

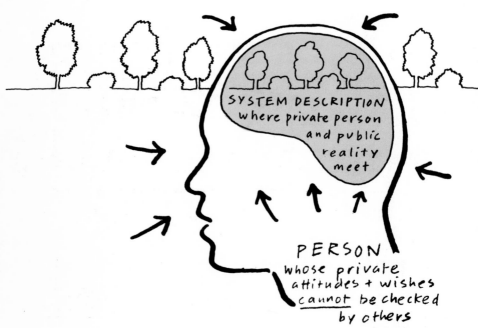

When different people, with different interests, prepare a systems description of the same situation they may generate very different pictures.

So when you talk about a particular systems description, you must show:

—the name of the system

—the person (or people) who own the description and named it

—what their special interest was in describing the system as they have done.

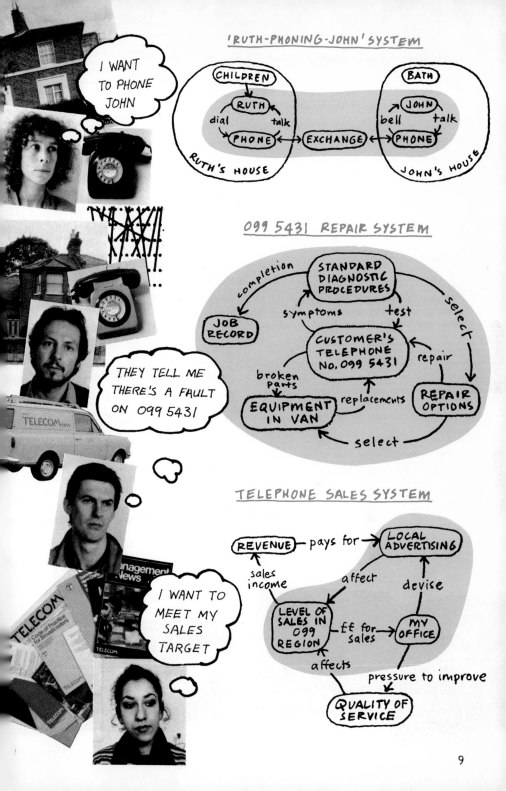

9

Levels of analysis and resolution

There is a hierarchy of possible levels of systems description.

For instance a systems description of a regional telephone sales strategy would be high on the hierarchy, with:

Broad scope (marketing information, demographic factors, labour agreements, etc.)

Coarse resolution (elements such as: monthly sales rates, current charges, etc.)

Long time scale ('structure' includes things like: cable networks; 'process' including things like: monthly trends in installation rates.)

Whereas a system description of 'Ruth phoning John' would be lower down the hierarchy, with:

Limited scope (one telephone connection)

Finer resolution (elements such as: Ruth, John, handsets)

Shorter time scale ('structure' includes the phone installation that would be 'process' for regional sales strategy; 'process' lasts one phone call).

Element

level for

Element   Sub-system

level for telephone repair

Element   Sub-system   System

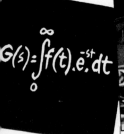

$$G(s) = \int_0^\infty f(t).e^{-st}.dt$$

level for British Telecom long-term planning

| Element | Sub-system | System | Environment |

level for Regional Sales Strategy

| Sub-system | System | Environment |

Ruth-phoning-John'

| System | Environment |

Environment

Types of system

Even within one particular system there may be sub-systems, or associated systems, of different types. For instance, in the telephone example you might find:

—Natural systems, such as the ecosystems of rats, or the weather systems of the wind, which may, incidentally, influence the telephone system by gnawing cables, or blowing down wires.

—Abstract systems, such as a set of linked mathematical equations or a computer programme. In the telephone example, these might be found in say, planning models or computerised invoicing systems.

—Designed systems, such as the telephone hardware itself.

—Systems of human activities (Ruth phoning, engineer repairing). These are what this book is mainly about.

You also have to decide whether a particular system or sub-system is to be treated as 'soft' or 'hard':

This 'car system' is **HARD**. Its properties are known. An engineer could predict its behaviour quite accurately.

oft' systems descriptions can be discussed and explored, but do not
tempt to represent the original situation precisely and
ambiguously because they involve emotional reactions, personal
lues and attitudes and shifting expectations. 'Soft' systems
scriptions are 'personal' rather than 'technical' in attitude, and tend
 be used most for 'people' systems, though some non-people
stems need to be treated in a 'soft' way too — e.g. some complex
d unpredictable machines, or intricate ecosystems.

e 'hard' systems style is precise, well-defined, and quantitative. It is
ed in situations where it makes sense to measure them, make
odels of them, and expect them to behave with a predictable
gree of regularity. Mechanical or electronic assemblies are often
efully described as 'hard' systems; highly routinised human activities
n sometimes be treated in this way.

When the tourists want to decide
what to do next, they talk about it;
they do not get out their calculators.
This 'holiday-decision-making system'
is SOFT.

So some systems are more tangible than others . . .

Some systems descriptions are so subjective and idiosyncratic that if someone else looked at the same components they would not recognise any systemic interconnections.

For instance, a regional sales representative may 'own' a human activity system for selling products in the S.W. Midlands. It may contain many objective components: a car, travel times by various routes, useful hotels, lists of contacts, etc. But their only connection is in the mind of the rep. They function as a coherent system only because this particular rep makes them do so, by using the car to drive down the routes, to go to the hotel, to meet the contacts, to . . .

There are other systems descriptions that, though still highly subjective, are rather more predictable in practice, given their owner's culture or social role. For instance, a monk, a civil servant, and an industrialist may well agree on low level technical descriptions (e.g. a system for cooking porridge) but when considering high level problems, such as how to govern the country their system descriptions and the actions that follow from them may differ dramatically, because:

They have different ultimate goals (pursuing enlightenment, implementing government policies, entrepreneurial development) and different criteria as to what short-term stresses are acceptable en route.

They have different theories about how events affect one another, in areas of great complexity where many theories are equally plausible.

They have such different uses of language that constructive discussion between them is very difficult, even though they are all responding to the same underlying human nature.

However some systems descriptions are much less personal and rely on more tangible links. In these cases, different observers are more likely to come up with quite similar descriptions. People who live in a village may be more closely connected to one another than to outsiders in terms of communication, friendships, employment, and many other factors. So very different systems description owners, with very different descriptive purposes, might still find themselves drawing quite similar systems boundaries around the village system.

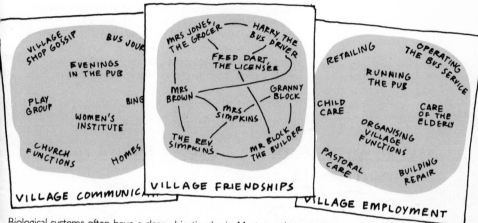

VILLAGE COMMUNIC...   VILLAGE FRIENDSHIPS   VILLAGE EMPLOYMENT

Biological systems often have a clear objective basis. Most people will see me as much the same person I was twenty years ago, even though every molecule in my body has changed many times. I am like the whirl in a bath plug-hole — the whirl remains, as the water flows through it. When I die, my body can no longer use energy to organise and maintain the recognisable pattern of 'me' and it disintegrates. Designed technical systems such as bridges or machines are usually deliberately made so as to present a single clear-cut definition to most observers.

Of course there are some very 'real' situations, (and some parts of every situation) where systemic (or any other) analysis is not possible because the situation is so complex and fast changing as to be incomprehensible. Then you talk of 'being overwhelmed', 'trusting to fate', or perhaps (if you are lucky!) 'riding a wave of good fortune'. Apart from devising emergency measures, or including prudent safety margins, all you can do about it is to let yourself be carried along, and try to stay afloat.

What holds a system together? Control.

Any persisting pattern of activity that can be described as a system must involve processes that hold it together; otherwise it would tend to degenerate. So the <u>structure and process</u> of a system and the <u>control</u> of the system are two sides of the same coin.

There are different types of control.

1. The natural ecosystems of tropical rain forests illustrate the most basic form of control. They can be stable for milennia, with no sens of purpose, no special controller, no free choice, no grand design. They are controlled by an immensely complex <u>self-maintaining causa network</u> that will hold itself in the same general state indefinitely, unless it is radically destabilised (as in human forest clearance schemes) by changes too drastic for the network to absorb. Here are some other much simpler networks that enhance or resist change:

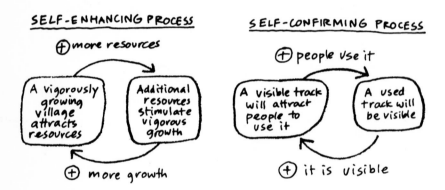

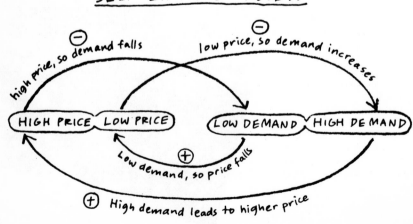

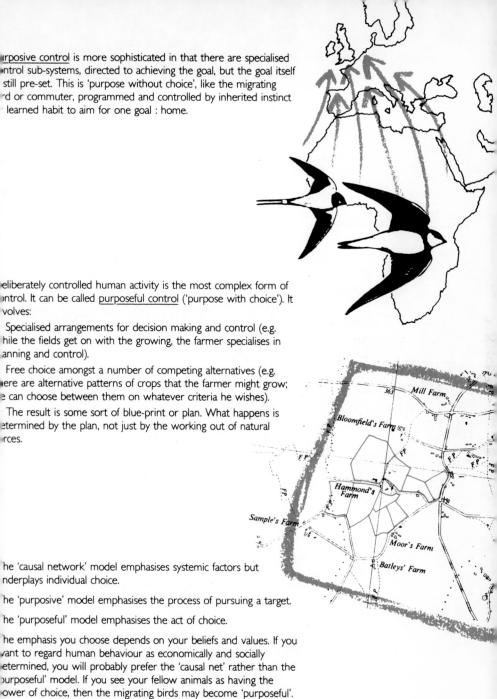

...rposive control is more sophisticated in that there are specialised ...ntrol sub-systems, directed to achieving the goal, but the goal itself ...still pre-set. This is 'purpose without choice', like the migrating ...rd or commuter, programmed and controlled by inherited instinct ...learned habit to aim for one goal : home.

...eliberately controlled human activity is the most complex form of ...ontrol. It can be called purposeful control ('purpose with choice'). It ...volves:

...Specialised arrangements for decision making and control (e.g. ...hile the fields get on with the growing, the farmer specialises in ...anning and control).

...Free choice amongst a number of competing alternatives (e.g. ...ere are alternative patterns of crops that the farmer might grow; ...e can choose between them on whatever criteria he wishes).

...The result is some sort of blue-print or plan. What happens is ...etermined by the plan, not just by the working out of natural ...rces.

...he 'causal network' model emphasises systemic factors but ...nderplays individual choice.

...he 'purposive' model emphasises the process of pursuing a target.

...he 'purposeful' model emphasises the act of choice.

...he emphasis you choose depends on your beliefs and values. If you ...vant to regard human behaviour as economically and socially ...etermined, you will probably prefer the 'causal net' rather than the ...urposeful' model. If you see your fellow animals as having the ...ower of choice, then the migrating birds may become 'purposeful'. ...may be as much a matter of respect and affection as of intellectual ...rgument.

Steering towards a target

Both purposive and purposeful models involve a <u>controller</u>. But it is not enough just to operate the control knob. You also need to check what happens when you do so — things may not work out as you expect!

This is <u>adaptive control</u>, sometimes called 'feed-back' or 'closed loop' control, because information about the results is fed back to the controller, so 'closing the loop': feel the water; if necessary take corrective action; feel the water again; corrective action again if needed; feel . . .

ADAPTIVE CONTROL

If the checking side of the loop is missed out, the controller won't know if something went wrong. But checking and adjusting can be expensive; and if the time-lag in the closed loop means that the adjustment is too out of date when you make it, feed-back can even make things <u>worse.</u>

So sometimes we have <u>non-adaptive control</u> where most of the effort goes into setting the system up correctly and reliably in advance, so that subsequent checking is not needed (though if anything does go wrong, you will probably get to hear about it sooner or later through critical reactions, or serious mishaps: chickens come home to roost eventually!). For instance, we checked the proofs of this book carefully, but once it was on the press no one proof-read each copy (until you did!). This is sometimes called 'feed-forward' or 'open-loop' control, because the controller has to predict in advance the exact action needed, and the loop is not closed by checking the results.

NON-ADAPTIVE CONTROL

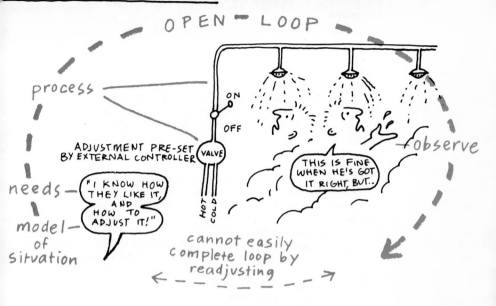

Both adaptive and non-adaptive control need:

Knowledge about the system — a 'model' that lets the controller judge the likely effects of possible actions.

Awareness of the 'needs' that the controller is trying to achieve.

Of course in a complex system there has to be some sort of overall coordinated control, as well as the control of individual tasks.

It's not too hard to see who made the purposeful choices in this system, and who got stuck with the purposive model:

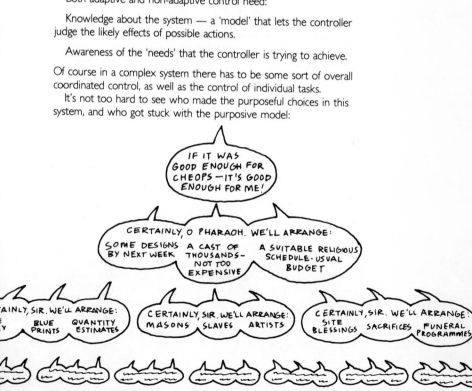

But where do needs come from? Objectives, goals and viewpoints.

The nitty-gritty short-term <u>objectives</u> ("I must fix the washing machine this weekend") and the <u>tactics</u> to achieve them ("This time it's going to be a D.I.Y. job") come from the practical needs of the local situation. Objectives are not very subjective!

The local situations may arise from broader, longer term, <u>goals</u> ("I've always wanted a nice house with lots of labour-saving aids") and the <u>strategic plan</u> to achieve them ("secure employment, a mortgage within my means and a family to generate the labour that needs to be saved"). There's a much stronger subjective element at this level — someone else might find that dream a nightmare ("gadget-ridden suburbia").

The goals may in turn arise from even broader policies or principles, such as a country's constitution, a committee's terms of reference, or a particular ideological analysis of why things are as they are ("People get what they deserve, so you'll have to work hard; then you'll get that house").

But these successively broader and more subjective guidelines become harder and harder to convey in words, eventually becoming a barely conscious blend of cultural pressures, and personal attitudes, wishes and perceptions. This forms the '<u>appreciative system</u>' that generates your values and priorities, and the '<u>weltanschauung</u>' (world-view) that structures the way that you see things.

We often find it hard to recognise these very broad influences at work in ourselves until we meet someone with a very different world-view and find we cannot see eye to eye, every attempt to communicate leaving us more confused and frustrated.

Such factors are often at work in a more positive way when we use phrases like:

—an action that 'looks good' or 'feels wrong'

—feeling your attention drawn to some issue.

—a hunch or 'gut reaction'.

—feeling that 'the pieces are coming together'

Where we organise ourselves together we have to find ways of harmonising the many individual world-views to coordinate our goals and objectives. For instance:

1. People may converge on one particular shared direction just by limiting their options; for example:

   by knowingly or unknowingly accepting a common culture of shared norms

   by letting someone else tell them what to do: leader, teacher, manager, expert, parent . . .

2. Sometimes there is an attempt to synthesize a genuinely shared direction; for example:

   by some political procedure, such as a debate followed by a majority vote

   by working for mutual understanding so as to maximise consensus and avoid spurious conflict

   by negotiation and bargaining

3. A third option is to find a quantitative way of expressing the aspects of particular actions that matter to us and what they are worth to us. In theory, you can then calculate rationally the 'best' mix of pros and cons. But putting numbers to subjective feelings can be very deceptive, and 'because-the-computer-says-so' is not always a reliable way of promoting shared commitment.

So who are the people with all the view-points?

A number of roles recur in many systemic analyses (or indeed in any other form of organisational consultancy). For instance:

The agent (or agency): The person, or group of people, trying to get some kind of constructive action going — perhaps as problem-solvers, advice givers, change facilitators, or designers. If you are reading this book, you are probably thinking of yourself in this role.

The client, who commissions the agent and pays the fee.

The problem-owner, in whose 'patch' the problem or opportunity is focused.

The gate-keepers who control channels or operations on which action will depend.

The power figures, whose authority, influence, or resources will prevent or allow the action.

Other participants within the system who keep it going.

A wide range of roles that play an active part in the system's environment — suppliers, users, customers, enquirers, transporters, message carriers, . . .

Other roles in the system's environment where consent is needed — local voters, regulatory officials, critics and objectors, the legal system . . .

And many others, sometimes with every role in a different person, sometimes with a few people each wearing many hats.

A human activity system, like a theatrical performance or a novel, represents a human drama; the stock characters of the human condition are as likely to appear in it as in any other descriptive medium, and the criteria you would apply to test the truthfulness and communicative power of a drama or novel apply just as much to systems description.

Why should any one else believe descriptions that can be so personal?

Physical sciences judge 'truth' or 'validity' by how well a descriptive model can predict how things will turn out under a wide range of conditions, on different occasions, and for different observers.

In the social sciences, situations are usually so complex and variable that they never repeat exactly. And because we are 'free' beings, we can choose (within limits) what we do; so we can often prevent a predicted outcome happening if we want to — we have the ability to subvert predictions. In applied social sciences, as when we apply systems ideas to organisational analysis, there are additional problems due to deadlines and limited resources, and the central involvement of both client and analyst. So unambiguous objective validity, as understood by the physical sciences, is impossible, and probably meaningless, in our context.

Here are some more useful criteria:

A good systemic model describes a situation to its participants in such a way that:

— they feel that it makes sense of their experience of the situation and its context.

— they can commit themselves to it as a framework around which to coordinate their actions.

— when they do so they find it useful and realistic so that their expectations of it are appropriate and they are unlikely to be surprised by unexpected outcomes.

Descriptions that meet these criteria will often be rational and objective in style. But they don't have to be. For instance the beautiful stories of the two thousand year old Indian 'Tales of Bidpai' in which King Dabschelim learns about diplomacy, human relations, and government by listening to an intricate network of animal fables, each exploring a particular pattern of relationships and outcomes, are more fun than the average twentieth-century management text, wiser than most, have sold more copies over the milennia, and can meet the criteria!

A checklist of 'good practice':

1. Many details of the description will depend on the analyst's judgement, which the participants will have to take on trust. So they must be able to make their own judgements of the analyst's skill and trustworthiness from track record, personal qualities, etc.

2. All data must be 'quality controlled' to minimise artefacts and errors.

3. Data collection must be thorough, broad, and impartial leaving each participant feeling that the analyst has collected the data that seem relevant from their viewpoint. There should be no reason to believe that potentially relevant information has been carelessly bypassed or deliberately concealed.

4. There should be evidence that the data collected influenced the description! Collection must not be merely a public relations exercise, and thinking must evolve as information comes in.

Nevertheless, the final selection and interpretation of the data for the description will inevitably reflect some viewpoints more than others — eg those of the initial terms of reference. This should be made explicit, so that constructive debate is made easier.

Participants must be able to engage fully with the description; the same description may 'work' on one setting, and be useless in another. So:

— it must be in language the participants can use, no longer or more complex than they can grasp in the time available, and able to hold their attention and interest.

— it must suggest actions that they see as feasible and helpful, within the scope of their authority, and within their tolerance of change.

— they must be able to 'own' it intellectually and emotionally as an accurate and acceptable representation of their experience (which often means that they should have been closely involved in producing it).

— if it calls for a radical rethink, people have to be emotionally as well as intellectually ready for it. If they are, you may get the emotional 'moment of truth' or sudden: 'aha'. If badly judged it may be passively ignored because it still seems irrelevant, or actively rejected because it provokes anxiety, anger or frustrated boredom.

Stages in the journey

Practical change involves a particular kind of systems drama:

Finding some practical route from *here* — the present situation and all the messy practical constraints, fears and opportunities that it involves . . .

And making the mental and phsyical journey to *there* — some future state that we envisage, perhaps in clear detail, or perhaps as little more than 'getting away from here'.

HERE ————————————> THERE

But few dramas come to a happy conclusion in the first act, and first time around it is unlikely that you will find exactly the right compromise that meets real needs, that can be achieved effectively, and that is actively supported by the people involved. Yet if you get it wrong in a very public way, the response may prevent any 'second chance.'

So usually you try not to jump in at the deep end, and keep most of the trial-and-error off-stage, in some protected setting where you don't court disaster every time you take a risk, making it useful to distinguish between:

FLUID

The *fluid* 'bright ideas' phases in which you can change your mind without serious consequences — you can toy with possibilities, find out how things tick, explore the potentialities of a situation, dream up alternative proposals, sketch ideas, prepare pictures and descriptions, experiment with models, compare options.

And the *consolidated* 'real world' phases, which are hard to reverse, so that changes have very serious consequences — doing it for real, negotiating with important people, getting authoritative approval, facing the real costs of stopping or starting things; competing for limited resources, accepting legally binding contracts, and taking real responsibility for harmful consequences.

CONSOLIDATED

Four styles of work

So the journey is not only from *here* to *there*, but also from *fluid* to *consolidated*:

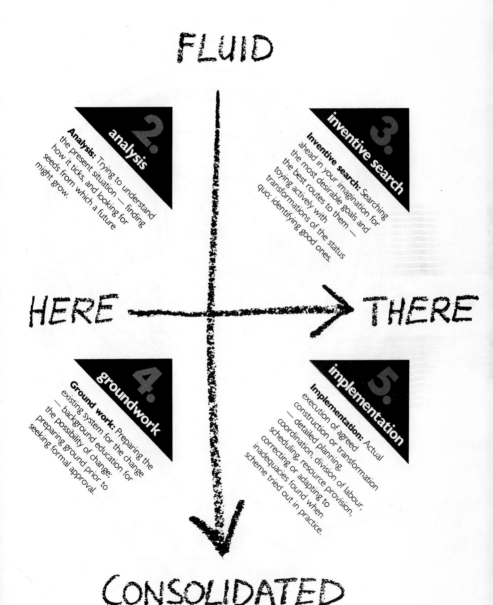

FLUID

2. analysis

Analysis: Trying to understand the present situation — finding how it ticks and looking for seeds from which a future might grow.

3. inventive search

Inventive search: Searching ahead in your imagination for the most desirable goals and the best routes to them — toying actively with transformations of the status quo; identifying good ones.

HERE ———→ THERE

4. groundwork

Ground work: Preparing the existing system for the change — background education for the possibility of change; preparing ground prior to seeking formal approval.

5. implementation

Implementation: Actual execution of agreed construction or transformation — detailed planning coordination, division of labour, scheduling, resource provision; correcting or adapting to inadequacies found when scheme tried out in practice.

CONSOLIDATED

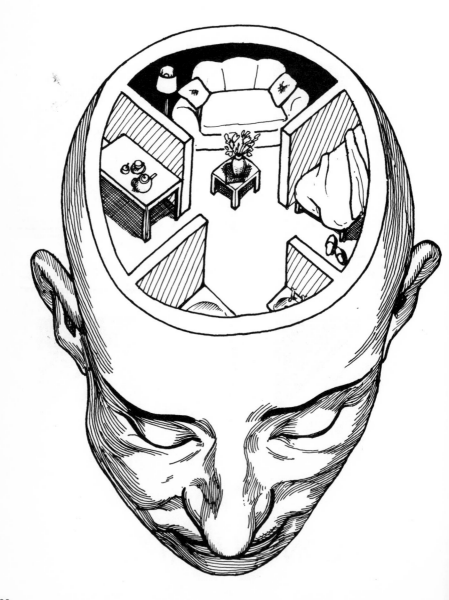

When to work in which style

Much of the skill in effective action lies in knowing how best to mix these four styles of work, and quite a bit has been written on how to do it (see 'Further Reading' on page 108).

You can think of the four styles as rather like rooms in a four-roomed house. Living doesn't have a fixed sequence. We like to move easily from one room to another when we need to. All the rooms work together as a 'system for living in'. It is quite normal to spend a lot of time in one, or pop briefly into another; but if we spend all our time frantically running from one room to another, or alternatively lock ourselves all the time in one room only, something has probably gone wrong.

Just as each room in a house imposes its own personal role, atmosphere, functions, and customs, so each of our styles of work tends to have;

— its own specialists

— its own approach and driving force.

— and its own typical problems and techniques.

The logical sequence is of course from *fluid* to *consolidated* and from *here* to *there*, giving us:

Analysis ⟶ Inventive search ⟶ Groundwork ⟶ Implementation

But it is rarely as simple as that, and Implementation is rarely the 'last word'. All sorts of sequences can occur for instance:

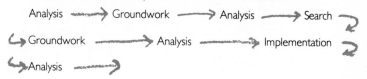

Analysis ⟶ Groundwork ⟶ Analysis ⟶ Search ⟶

↳ Groundwork ⟶ Analysis ⟶ Implementation ⟶

↳ Analysis ⟶

. . . but analysis is as good a place to start as any:—

Analysis is needed for unfamiliar situations. You are no longer sure what to do.

You could just "jump in at the deep end", learning by random trial and error and intuitive hunches — the heuristic approach.

But a practically oriented <u>description</u> or <u>map</u> of the situation can often be very helpful.

Description makes a situation thinkable, and that makes it easier to talk about, to share, and to act within.

Some descriptions are more useful than others. But if you can get
e right description, the action needed will often be obvious.

A problem is 'solved' when the people involved have found an
tionable way of describing it.

But before you roll up your sleeves and start describing, there are
me useful preliminary activities to bear in mind. You don't want to
me in too soon on what may turn out, much later, to have been
e wrong area.

This man's problem presented itself in an all too direct and
rceful way! But in a more complicated situation, it is often harder
see the wood for the trees.

Some stages in building up a better vantage point . . .

first, a preview:

LOTS OF ITERATIVE
TO-ING AND FRO-ING

1 Initial _Awareness_ of some aspects of a situation that hold your attention and lead you to explore them.

2 _Commitment_ to a particular task develops as you become able to state what it is, and why you are interested.

DISCARDED TASKS

3 You then _Test_ your chosen Task to see if systemic analysis would be a sensible next step.

4 To start systemic analysis, begin _Separating_ out different areas of the situation into a range of more or less relevant systems.

5 Then _Select out_ which one or two of these to concentrate your limited analytic resources on for the time being.

6 And that leaves you ready for _Detailed Description_ . . .

DISCARDED SYSTEMS

. . . and eventually, perhaps, new design and implementation . . .

. . . and now a closer look at each step:

Step 1: The origins of awareness

You don't quite know how the discussion started, but it did. People kept on tossing in images, viewpoints, experiences, information . . .

An ambiguous but rich and complicated picture began to form in your minds. Again and again strongly expressed views tempted you to jump prematurely to 'short-cut' conclusions, and you had to challenge yourselves with yet other viewpoints to stop your system of ideas becoming closed.

The emerging picture is not at all tidy. It is very loosely and inconsistently structured and full of loose ends — a hotch-potch of facts and illusions, and of symbols and images that each say particular things quite clearly, but hint at many others. There are all sorts of different angles: human, technical, financial, political, organisational, aesthetic and atmospheric, procedural, motivational . . . It is a resource-heap of largely unsorted raw material that has seemed relevant as the themes of the discussion have evolved. Some bits of the heap are quite organised; others are still chaotic.

You are trying to develop a mental sketch map of the 'territory' you are becoming interested in. While it is the <u>mental</u> sketch map that matters most directly in guiding your thinking, it often helps if you also develop a <u>physical</u> picture on, say, a blackboard. This is often called a 'situation summary' or 'rich picture'. There is nothing fixed about a picture like this. It is never 'finished' (though you may eventually abandon it when its usefulness is over) — it adds new material, discards old material, reorganises and develops all the time, focusing your discussion and guiding your enquiries.

At one point it might look like this:

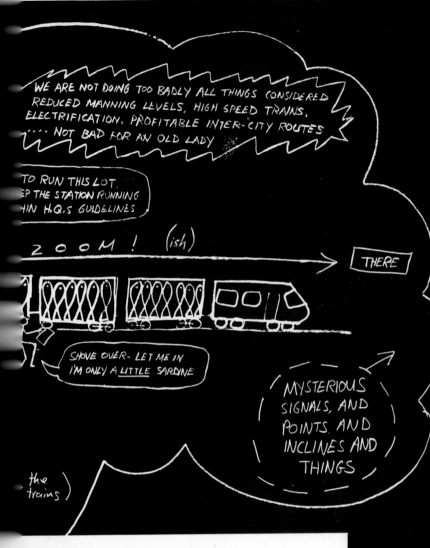

Step 2: Converging towards commitment

As the picture begins to take shape, so particular issues take on a sharper focus. There are lots of possible issues. You can't solve everything at once. Which should you tackle first? Perhaps you would like to make a short-list of issues you might commit yourself to. Check them out on the next step before homing in on one stated commitment . . .

Step 3: <u>Testing</u> for systemic content . . .

Commitment statement	Is it reasonably clear from the statement what 'success' in this task would consist of?	Is the extra effort spent on analysis justifiable? If the task is simple and well understood, it may not be.
STATION MASTER: "I am responsible for keeping this station running smoothly within BR's guidelines — so I need to understand it."	✓	✓
OLD AGE PENSIONER: "I feel frightened and confused by places like this. Someone should do something to make it easier for people like me."	✓	✓
SYSTEMS STUDENT: "I've been told to make a systemic description of this, to practice the technique."	✓	✓
TOURIST WITH CAMERA: "Gee, I just want to capture the atmosphere of this great old station!"	✓	✓
LITTLE BOY: "I want to see if I can collect forty different train numbers today!"	✓	No — Enjoy yours
DOWN-&-OUT: "I'm just sitting here, waiting and watching."	Not clear yet	?

Is the analysis to be an investigative and functional one — e.g. 'how does it work' or 'what causes this'?	Is the particular nature of this situation important for the particular task you have chosen?	**Verdict:** systemic content	Notes
✓	✓	YES	IDEAL CASE
✓	✓	YES	Complaints and direct expressions of need are often good starting points.
✓	NO — any situation would do for practising on	✗	Why choose railways?
Aesthetic judgement rather than functional analysis	✓	✗	Not a systemic issue.
No	✓	✗	Needs action, not analysis.
?	?	Can't tell	Back to AWARENESS or COMMITMENT stages.

The Task: "I feel frightened and confused by places like this. Someone should do something to make it easier for people like me". (O.A.P.)

Having chosen The Task you are ready to start <u>separating</u> out the component systems, so . . .

Step 4: Separation ▼

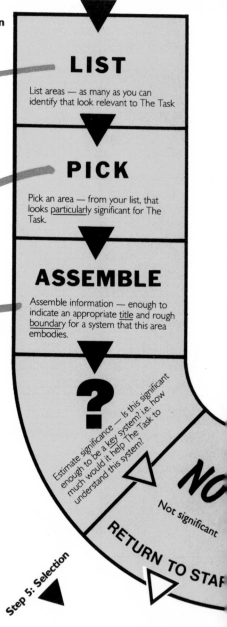

LIST

List areas — as many as you can identify that look relevant to The Task

B.R. complaints procedure

B.R. marketing

Attitudes to violence and crime

Training of station staff

Agencies that act for the old and infirm

Family and friends

PICK

Pick an area — from your list, that looks <u>particularly</u> significant for The Task.

ASSEMBLE

Assemble information — enough to indicate an appropriate <u>title</u> and rough <u>boundary</u> for a system that this area embodies.

Title: 'A system to ensure that relevant decision-makers pay more attention to the needs of the old and infirm'.
Typical components within boundary: Age Concern, Social Services, local councillors, M.P.s, rail transport user groups, research programmes, government grants and charity collections.

Typical components within the environment: Government policy, B.R. finances, competing B.R. priorities and decision making, newspapers, radio & T.V.

?

Estimate significance — Is this significant enough to be a key system? i.e. how much would it help The Task to understand this system?

NO

Not significant

RETURN TO START

Step 5: Selection ▼

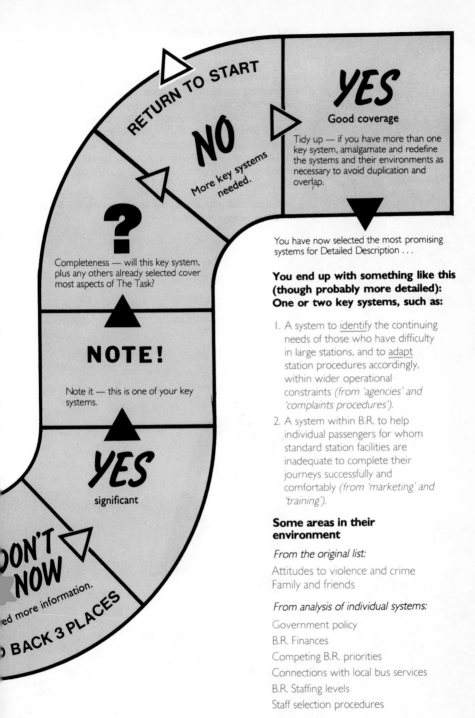

RETURN TO START

NO

More key systems needed.

YES

Good coverage

Tidy up — if you have more than one key system, amalgamate and redefine the systems and their environments as necessary to avoid duplication and overlap.

?

Completeness — will this key system, plus any others already selected cover most aspects of The Task?

NOTE!

Note it — this is one of your key systems.

YES

significant

DON'T NOW

...ed more information.

...O BACK 3 PLACES

You have now selected the most promising systems for Detailed Description . . .

You end up with something like this (though probably more detailed): One or two key systems, such as:

1. A system to <u>identify</u> the continuing needs of those who have difficulty in large stations, and to <u>adapt</u> station procedures accordingly, within wider operational constraints *(from 'agencies' and 'complaints procedures')*.

2. A system within B.R. to help individual passengers for whom standard station facilities are inadequate to complete their journeys successfully and comfortably *(from 'marketing' and 'training')*.

Some areas in their environment

From the original list:

Attitudes to violence and crime
Family and friends

From analysis of individual systems:

Government policy
B.R. Finances
Competing B.R. priorities
Connections with local bus services
B.R. Staffing levels
Staff selection procedures

43

Step 6. Detailed Description: an introduction

'Description' can mean anything from a child's essay on 'What I did on my holidays', to a computer simulation of the national economy, or the atomic theory of matter. All represent attempts to make various experiences communicable.

Analysts tend to use the word 'model' rather than 'description' when they want to indicate that the description has been prepared in a careful 'quality controlled' way rather than by informal verbal description, though the distinction is far from absolute.

Whether they are systemic or not, description and modelling always involve simplification, because you have to simplify in order to communicate.

Sometimes you are trying to communicate what you saw or experienced

Leonardo modelled reality using a wide range of pigments and different kinds of brush-strokes, and the full spatial possibilities of a 2-dimensional surface.

← But his description is still quite recognisable, even after gross simplification...

← and for police identification purpos the lady can be redu to a selection from small stock of eyes, noses, mouths, etc...

..and for stretch fab design, she's just one of thre standar size

70% COTTON
30% POLYESTER S DN
STER M

HAND WASH ONLY
COOL (30C)
COLD RINSE - SHORT SPIN
DO NOT HAND WRING

Alternatively, you may be trying to communicate more abstract thoughts about reality

This mediaeval cosmology thought of man as the most important being, on the most important place (the earth), at the centre of the concentric orbits of the planets and stars. God pervaded everything and held it together. If you accepted this model, it supported a certain kind of power relationship between God's Church, and Man's State.

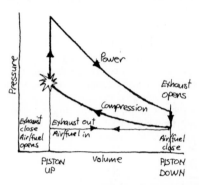

This tells me about the major events in a four stroke petrol engine. But it doesnt help me to find out why my car didn't start this morning.

The Market Economy made Simple: models of three political viewpoints.

The capitalist, 'free market', view: if you leave the market alone, the economy will self-correct to its natural equilibrium.

The mixed-economy, 'interventionist' model: it is self-righting to small changes. But the State must intervene if it goes too far off-course.

The 'centrally-planned', Marxist, view: the economy, left alone, is totally unstable. Only with State control will it hold up.

Each of these views is a different, but quite defendable, simplification of the observable data but they reflect different values and priorities, often imply different actions, and are rallying points for different people.

Systemic models also involve simplification, but <u>practical relevance</u> then provides the criterion for what to include or omit.

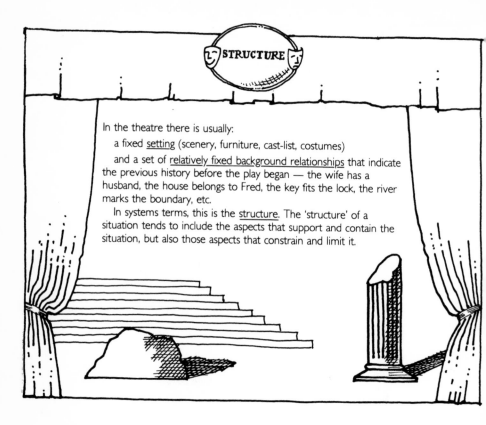

STRUCTURE

In the theatre there is usually:

a fixed <u>setting</u> (scenery, furniture, cast-list, costumes)

and a set of <u>relatively fixed background relationships</u> that indicate the previous history before the play began — the wife has a husband, the house belongs to Fred, the key fits the lock, the river marks the boundary, etc.

In systems terms, this is the <u>structure</u>. The 'structure' of a situation tends to include the aspects that support and contain the situation, but also those aspects that constrain and limit it.

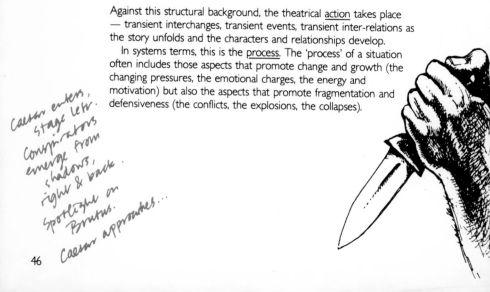

Against this structural background, the theatrical <u>action</u> takes place — transient interchanges, transient events, transient inter-relations as the story unfolds and the characters and relationships develop.

In systems terms, this is the <u>process</u>. The 'process' of a situation often includes those aspects that promote change and growth (the changing pressures, the emotional charges, the energy and motivation) but also the aspects that promote fragmentation and defensiveness (the conflicts, the explosions, the collapses).

Caesar enters, stage left. Conspirators emerge from shadows, right & back. Spotlight on Brutus. Caesar approaches...

You can also think of the difference between structure and process as like the difference between nouns and verbs. If 'Caesar rules the Roman Empire' Caesar and the Roman Empire are structure, ruling is process.

But verbs can become nouns. If 'Caesar is the ruler of the Roman Empire', 'the ruler' is now a noun — process has become structure.

So you have a lot of flexibility as to which parts you treat as noun-like structure, and which you treat as verb-like process. But you need to be clear about which you are using as which, you need to be consistent about it, and you need both structure and process — description won't work if it is all nouns or all verbs.

Structure and process are not just convenient descriptive categories. The relationship between structure and process often has a marked effect on the 'climate' of a situation, and since it involves the balance between the forces for stability and the forces for change, it may determine the directions that events may take:

Cas. Why, man, he doth bestride the
    narrow world
Like a Colossus, and we petty men
Walk under his huge legs, and peep about
To find ourselves dishonourable graves.
Men at some time are masters of their fates:
The fault, dear Brutus, is not in our stars,
But in ourselves, that we are underlings.

Relationships can be either structure or process depending on whether they are stable or transient. In its most general sense, two items have 'a relationship' if it makes a significant difference to one of them if the other is removed.

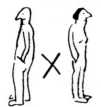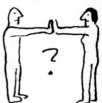

If one firm goes out of business, and this affects another firm, clearly the two firms 'had a relationship'. If one was the subsidiary of the other, this was probably a 'structural relationship'. If it is just that one had sold something to the other, but had not yet been paid, this is probably a 'process relationship'.

Relationships
can be
one-sided ...

Structure: Components and background relationships

Components are anything that people in the situation regard as
physical or abstract entities — structural features they treat in a
noun-like way — features that exist rather than happen.
For instance any of the following can be treated as 'components':

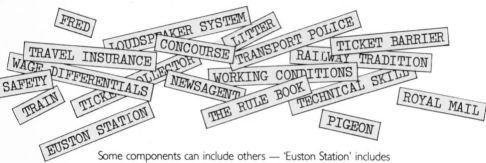

Some components can include others — 'Euston Station' includes
'concourse' which includes 'ticket barrier'. 'Fred' includes his
'technical skill'. 'Working conditions' include 'wage differentials' and
'safety'. In other words, there is a hierarchy of levels of components.
Background relationships are the relatively unchanging, stable,
relationships such as:

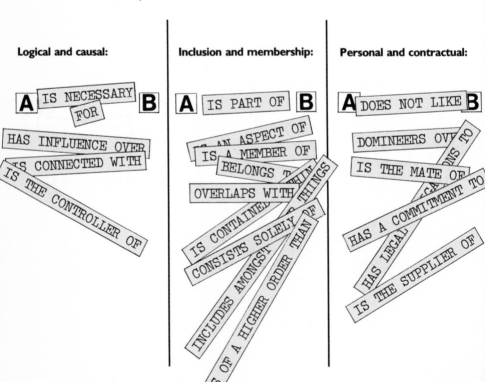

Representing the background relationships between components

1. The 'box and arrow network' (digraph) convention:

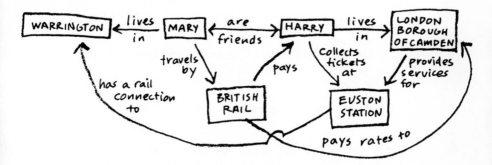

The digraph convention is good for showing the network of relationships that are all at one level of analysis, or are in a fully defined hierarchy.

But it is not good at showing, for instance, that the level of unemployment is different in Warrington and Camden, that this has general effects, but that it is hard to be sure of its particular effects on Harry or Mary.

We need to add in a second convention, that is much better at showing relationships of inclusion and membership and dealing with relationships on multiple levels, though less specific about particular details of relationships:

2. The 'overlapping and concentric areas' (Venn) convention:

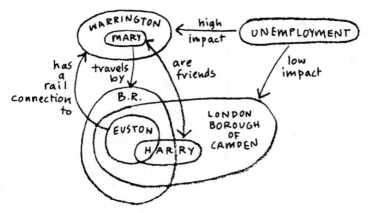

While some relationships can only be represented in one or other of these conventions, many can be represented in either, so that often you can choose the most convenient mixture for your diagram.

49

Process and transient relationships

While it is not too hard to represent the fixed <u>structure</u> of a situation as fixed lines on a diagram, capturing the changing, ephemeral, *process* is much more difficult — many of the really transient processes are bound to be lost in description — think how much a performer has to add to the bare words of a play script or a music score to make it 'come alive'.

The chronology of events is a part of the process. Here is one way of representing the story of the collision that occurred on 6th January 1968 between a train and a 162 ton transformer being transported by road. It was the first serious accident in the U.K. involving one of the 'half-barrier' level crossings that had recently been introduced.

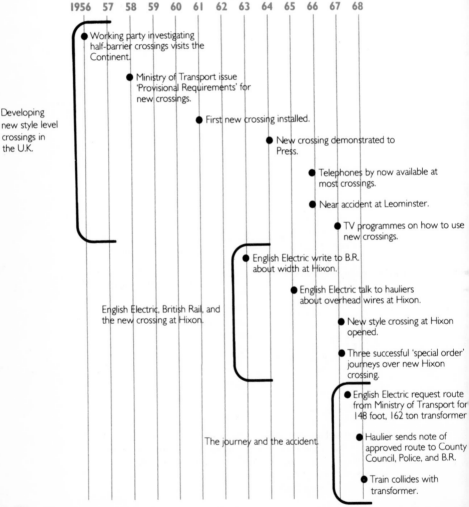

1956 57 58 59 60 61 62 63 64 65 66 67 68

Developing new style level crossings in the U.K.

● Working party investigating half-barrier crossings visits the Continent.

● Ministry of Transport issue 'Provisional Requirements' for new crossings.

● First new crossing installed.

● New crossing demonstrated to Press.

● Telephones by now available at most crossings.

● Near accident at Leominster.

● TV programmes on how to use new crossings.

English Electric, British Rail, and the new crossing at Hixon.

● English Electric write to B.R. about width at Hixon.

● English Electric talk to hauliers about overhead wires at Hixon.

● New style crossing at Hixon opened.

● Three successful 'special order' journeys over new Hixon crossing.

The journey and the accident.

● English Electric request route from Ministry of Transport for 148 foot, 162 ton transformer

● Haulier sends note of approved route to County Council, Police, and B.R.

● Train collides with transformer.

50

The inputs to, and outputs from, a structural component are often usefully thought of as part of the 'process':

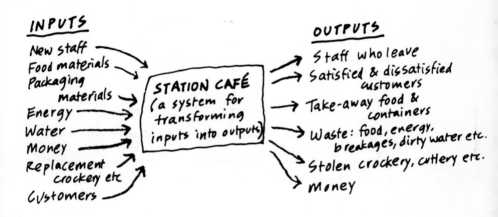

Here the idea has been developed to show the flows amongst a set of structural components:

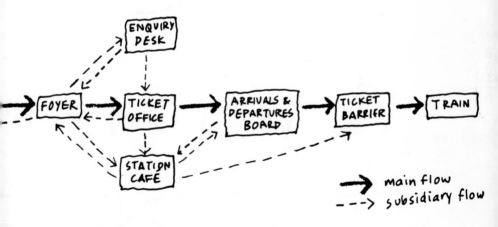

Process can be represented in an even purer form by showing the flows between functions rather than components (e.g. between: arrival, finding information, waiting, . . . rather than: foyer, enquiry desk, cafe, . . .). It is usually best to stick to either components or functions, rather than mixing them.

Another aspect of process is the mapping of consequences

This example, based on design problems that affected the new rail system for the San Francisco area, is a post hoc analysis of some unexpected <u>consequences</u> of adopting a particular broad policy.

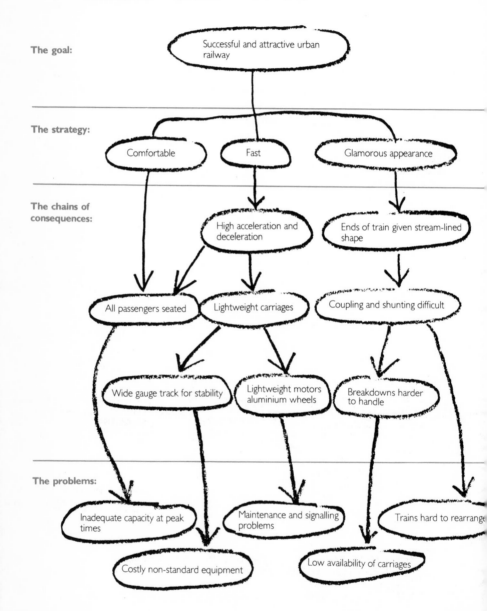

The goal:

Successful and attractive urban railway

The strategy:

Comfortable

Fast

Glamorous appearance

The chains of consequences:

High acceleration and deceleration

Ends of train given stream-lined shape

All passengers seated

Lightweight carriages

Coupling and shunting difficult

Wide gauge track for stability

Lightweight motors aluminium wheels

Breakdowns harder to handle

The problems:

Inadequate capacity at peak times

Maintenance and signalling problems

Trains hard to rearrange

Costly non-standard equipment

Low availability of carriages

This example, based on the Hixon incident, acts as a <u>fault tree</u> to identify possible precursors that might contribute to such an accident:

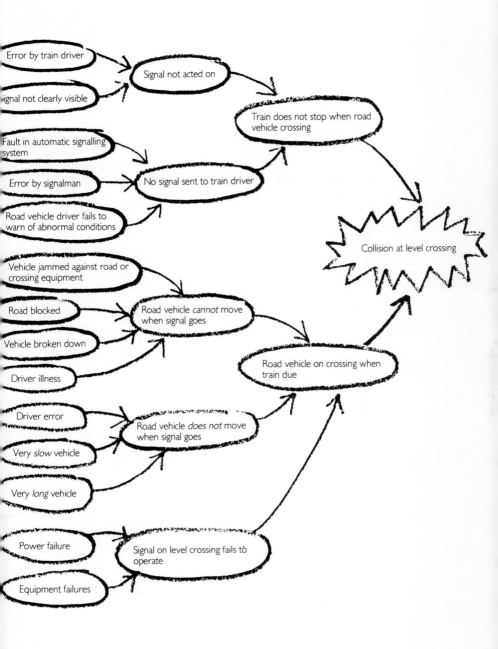

But within the analysis, there will often be problems . . .

In our supermarket 1 in 10 egg boxes can't be sold because of cracked eggs . . .

> **In our supermarket 1 in 10 egg boxes can't be sold because of cracked eggs . . .**

Is it really a problem?
- Some bakeries like pre-broken eggs
- Eggs are high in cholesterol, and involve keeping hens in battery cages— do we really want them?
- Whose problem is it anyway?
- Who does it affect?

A clearer view of the problem

After some discussion, it seems that we are talking about distribution from rural producers to small urban retailers, whose customers want unbroken eggs, in small quantities at a time. We are not concerned with bulk sales to bakeries or major national chains, or deliveries by middle-men.

Now we can start searching.
One way is to take an obvious lead,
and follow the train of thought from it,
exploring each successive step and
each possible branch.

In our supermarket 1 in 10 egg boxes can't be sold because of cracked eggs ...

We need better packing. Maybe thicker polystyrene

This will be more expensive. The cost of eggs may then be too high

Cheaper polystyrene?

Perhaps lower grade polystyrene

Perhaps use state subsidised supply from Hungary

But box may then be too weak

But this might bring supply and import problems.

Redesign box to compensate for weaker materials

Commission industrial design group for feasibility study

Working sequentially

This is the tidy, controlled, conscious kind of thinking that systematically explores each new step or alternative as it unfolds.

This type of working needs patience, time and care, and often depends on thorough background technical knowledge (e.g. where to find the polystyrene suppliers trade journal).

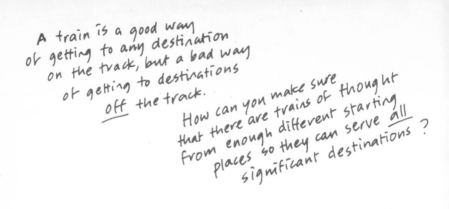

A train is a good way of getting to any destination on the track, but a bad way of getting to destinations *off* the track.

How can you make sure that there are trains of thought from enough different starting places so they can serve *all* significant destinations?

Perhaps we need the source of eggs nearer to retail outlets

Perhaps shells can be strengthened through selective breeding of hens

Perhaps we need to persuade people to like broken eggs

In our supermarket 1 in 10 egg boxes can't be sold because of cracked eggs . . .

Perhaps eggs can be radically re-packaged (without shells?)

Perhaps we need better transportation

Perhaps we could re-introduce powdered eggs.

We need better packing. Maybe thicker polystyrene

This will be more expensive. The cost of eggs may then be too high

Cheaper polystyrene?

Perhaps use state subsidised supply from Hungary

But this might bring supply and import problems.

Working for divergence

The techniques on the next few pages suggest ways of provoking yourself into attending to things you haven't been aware of, or have tended to ignore. Three general points:

Disrupting old habits creates anxiety. In the right setting and the right mood that feels adventurous and exciting. But it's very different if the setting is hostile and the anxiety is more than you can take.

The interplay of sequence and divergence often matters more than either on its own.

Ideas that feel weird or silly or alarming may not be viable solutions themselves. But if you can take them seriously enough to use them to start fresh trains of thought, they often lead to new and much more practical possibilities.

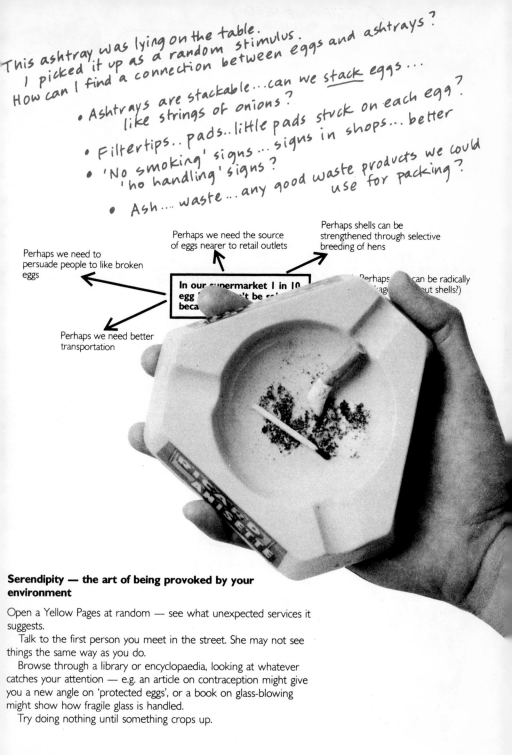

This ashtray was lying on the table.
I picked it up as a random stimulus.
How can I find a connection between eggs and ashtrays?

- Ashtrays are stackable...can we stack eggs... like strings of onions?
- Filtertips.. pads.. little pads stuck on each egg?
- 'No smoking' signs... signs in shops... better 'no handling' signs?
- Ash.... waste... any good waste products we could use for packing?

Perhaps we need to persuade people to like broken eggs

Perhaps we need the source of eggs nearer to retail outlets

Perhaps shells can be strengthened through selective breeding of hens

In our supermarket 1 in 10 egg ___'t be s__ beca___

Perhaps ___ can be radically ___ag___ ut shells?)

Perhaps we need better transportation

Serendipity — the art of being provoked by your environment

Open a Yellow Pages at random — see what unexpected services it suggests.

Talk to the first person you meet in the street. She may not see things the same way as you do.

Browse through a library or encyclopaedia, looking at whatever catches your attention — e.g. an article on contraception might give you a new angle on 'protected eggs', or a book on glass-blowing might show how fragile glass is handled.

Try doing nothing until something crops up.

Brainstorming — the art of being provoked by your friends

Find a place where you can be relaxed, noisy, private and uninterrupted.

Perhaps five or six of you — enough for a hubbub — not enough to lose anyone.

Total prohibition of criticism or 'put downs'; accept everything and where possible build on one another's ideas.

Make sure to write down everything that comes up.

Every now and then stop brainstorming, go back and see what you can do with what came up. Even the weirdest ideas may be staging posts to real progress. This review is crucial because it treats the fantasy with respect, but also brings it down to earth.

Health food enthusiast I want to know where my food comes from. I don't want me or my kids to be anaesthetised by musak and anonymous plastic egg boxes.

Egg marketing executive I'm determined never to go back to the days when you had to break each egg into a cup before you used it, because you never knew if it would be bad or not.

The farmer Battery farming and industrialised egg handling techniques are the only ways I can meet the demand for cheap eggs.

The advertising consultant Eggs sell because they are homely, clean, and easy to cook. The modern housewife likes the naturalness, but doesn't want a farm-yard in her kitchen.

The consumer I've got this cheap fridge made in Europe, and the egg compartment is too small for size 1 or 2 eggs . . .

Inviting criticism — the art of being provoked by your enemies

Find people who have different view points from yours.

It doesn't matter if they seem ill-informed, extreme, or politically unacceptable. What does matter is that you find some way of getting them to tell you what they think — or at least of imagining yourself in their shoes. Let yourself listen carefully.

By all means hold onto your own values, but don't throw out the baby with the bath water. Their different viewpoints may well raise issues and options you have not thought about very much, so they can be an important creative resource, if you will let them be.

Looking for new angles — the art of provoking yourself

If you can let your anarchic imagination have full freedom, while your analytic intelligence retains its full acuity, then each can support and provoke the other.

It is not easy to maintain this unstable balance. All too often the imagination tries to give you a holiday from your intelligence, or your intelligence tries to censor your unruly imagination.

Ask yourself:

Are there any other man-made situations with similar problems?

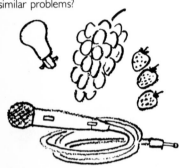

(Light bulbs,
microphones, fruit,
soft cheese, cartons
of milk ...)

Are there any similar situations in nature where small, fragile, objects get protected?

What are the fantasies you can spin around the
idea of an egg?

Let the idea of 'egg' rest in your mind for a few
moments, and then just reel off whatever strings of
words or images occur to you.

*(Egg, yolk, balloon, pop,
shell, war, egg-mobile,
brown & warm, sleep, womb,
snug...
   ...boiled egg, egg cup, china,
china ducks, float, water,
inflatable cushion...)*

Try to imagine what it would be like if you were as
small as an egg, and accompanied a batch of eggs
from their formation to their final use.

*(What happens? What frightens you?
How might you avoid injury?
If you set up an 'Egg Rights' campaign,
what would you be asking for?)*

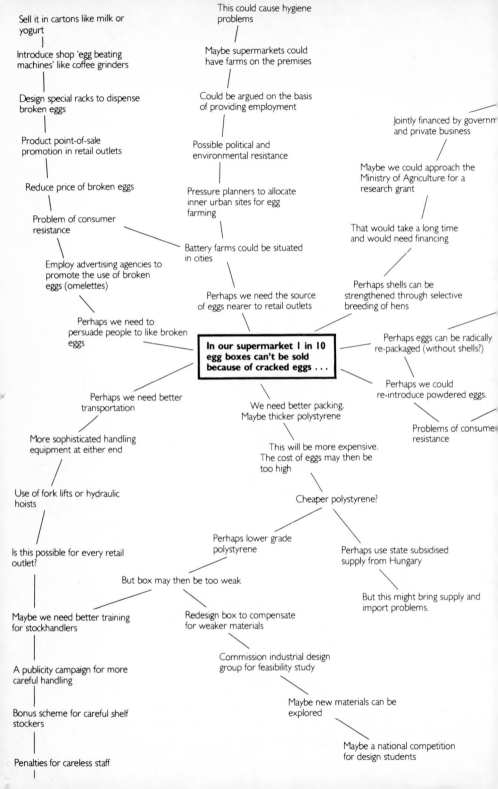

Sell it in cartons like milk or yogurt

Introduce shop 'egg beating machines' like coffee grinders

Design special racks to dispense broken eggs

Product point-of-sale promotion in retail outlets

Reduce price of broken eggs

Problem of consumer resistance

Employ advertising agencies to promote the use of broken eggs (omelettes)

Perhaps we need to persuade people to like broken eggs

Perhaps we need better transportation

More sophisticated handling equipment at either end

Use of fork lifts or hydraulic hoists

Is this possible for every retail outlet?

Maybe we need better training for stockhandlers

A publicity campaign for more careful handling

Bonus scheme for careful shelf stockers

Penalties for careless staff

This could cause hygiene problems

Maybe supermarkets could have farms on the premises

Could be argued on the basis of providing employment

Possible political and environmental resistance

Pressure planners to allocate inner urban sites for egg farming

Battery farms could be situated in cities

Perhaps we need the source of eggs nearer to retail outlets

In our supermarket 1 in 10 egg boxes can't be sold because of cracked eggs . . .

We need better packing. Maybe thicker polystyrene

This will be more expensive. The cost of eggs may then be too high

Cheaper polystyrene?

Perhaps lower grade polystyrene

But box may then be too weak

Redesign box to compensate for weaker materials

Commission industrial design group for feasibility study

Maybe new materials can be explored

Maybe a national competition for design students

Perhaps use state subsidised supply from Hungary

But this might bring supply and import problems.

Jointly financed by governm and private business

Maybe we could approach the Ministry of Agriculture for a research grant

That would take a long time and would need financing

Perhaps shells can be strengthened through selective breeding of hens

Perhaps eggs can be radically re-packaged (without shells?)

Perhaps we could re-introduce powdered eggs.

Problems of consume resistance

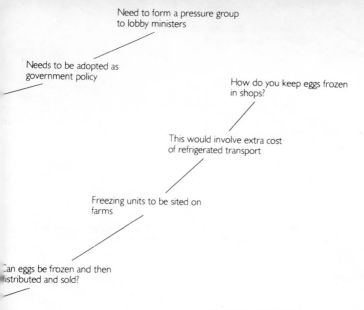

Need to form a pressure group
to lobby ministers

Needs to be adopted as
government policy

How do you keep eggs frozen
in shops?

This would involve extra cost
of refrigerated transport

Freezing units to be sited on
farms

Can eggs be frozen and then
distributed and sold?

Maybe it could be marketed
on the basis of war-time
nostalgia

Expansion by collecting all sorts of information . . .
and contraction to key elements on a situation summary.

Expansion by identifying many promising leads . . .
and contraction by following up only the best.

Expansion by formulating the problem in different ways . . .
and contraction by identifying the top priority version

Expansion by seeking many viewpoints . . .
and contraction by adopting one.

Expansion into the full complexity of a situation . . .
and contraction under one conceptual framework.

Searching for solutions is a cyclic process — many alternations
between wide expansion to get an overall perspective and to
develop, explore and map out the territory — and then contraction
to focus closely on the details of one particular aspect or option that
seems particularly appropriate for attention.

Like any form of iteration, multiple cycles of expansion and
contraction can be an expensive process. There is no virtue in them
for their own sake. But if the search is a productive one, there will
be real progress over the successive cycles, as a more tangible
outcome begins to define itself amongst the underlying confusion.

Expansion by generating all sorts of alternative plans . . .
and contraction by agreeing on one.

Expansion by advertising for many tenders for a job . . .
and contraction by selecting one.

Expansion by looking for many possible uses for a new facility . . .
and contraction by accepting particular uses.

Expansion by . . .

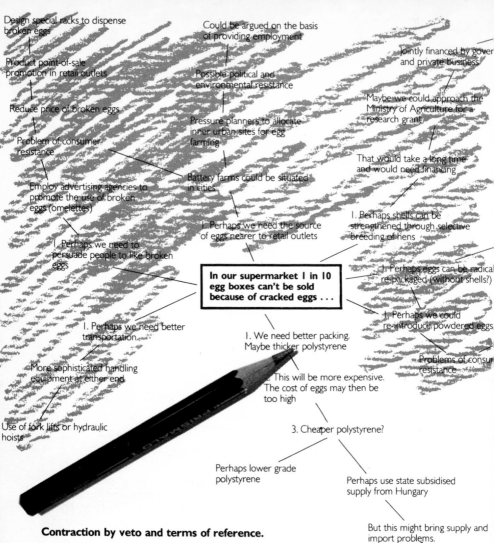

Design special racks to dispense broken eggs

Product point-of-sale promotion in retail outlets

Reduce price of broken eggs

Problem of consumer resistance

Employ advertising agencies to promote the use of broken eggs (omelettes)

1. Perhaps we need to persuade people to like broken eggs

Could be argued on the basis of providing employment

Possible political and environmental resistance

Pressure planners to allocate inner urban sites for egg farming

Battery farms could be situated in cities

1. Perhaps we need the source of eggs nearer to retail outlets

Jointly financed by gover and private business

Maybe we could approach the Ministry of Agriculture for a research grant

That would take a long time and would need financing

1. Perhaps shells can be strengthened through selective breeding of hens

In our supermarket 1 in 10 egg boxes can't be sold because of cracked eggs . . .

1. Perhaps eggs can be radical re-packaged (without shells?)

1. Perhaps we could re-introduce powdered eggs

Problems of consumer resistance

1. Perhaps we need better transportation

More sophisticated handling equipment at either end

Use of fork lifts or hydraulic hoists

1. We need better packing. Maybe thicker polystyrene

2. This will be more expensive. The cost of eggs may then be too high

3. Cheaper polystyrene?

Perhaps lower grade polystyrene

Perhaps use state subsidised supply from Hungary

But this might bring supply and import problems.

Contraction by veto and terms of reference.

If everyone were free to redesign everything, there would be no stability for anyone to do anything! Clear limits reduce anxiety and defensive manoeuvres. People need to have some idea of:

maximum likely cost

who will be affected or involved

the scope of your authority and competence.

But the reason why your clients need your help is because they don't know the answer — so their judgement of which lines of enquiry will be crucial may be suspect. And tight constraints are sometimes imposed in order to stop you rocking the boat.

So negotiating to maximise the scope for useful enquiry while safeguarding the organisation is a subtle, important, and ethically complex business.

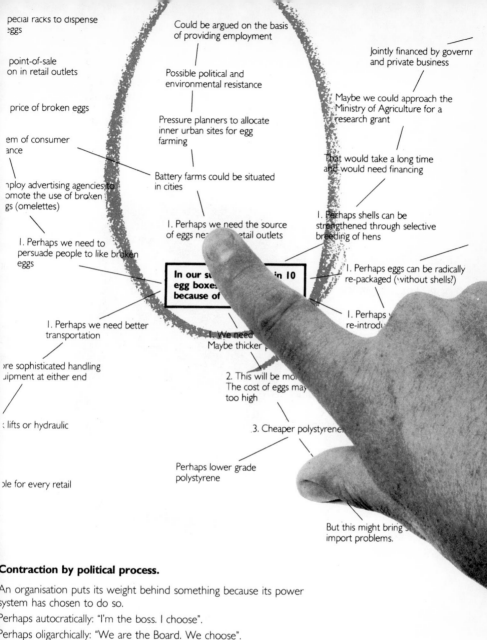

Contraction by political process.

An organisation puts its weight behind something because its power system has chosen to do so.

Perhaps autocratically: "I'm the boss. I choose".

Perhaps oligarchically: "We are the Board. We choose".

Perhaps democratically: "The feeling of the group is that we should choose that option".

Naturally such a choice will involve some element of skilled judgement and rational self-interest. But in a 'political' decision the roles of the people who made the choice, and the fact that they chose matter more than the rationality of the choice.

Contraction by rational optimising.

If you know exactly what the key characteristics of an ideal solution would look like (it must cost no more than £W, have a speed of at least X, be purchased from Y, fit into Z, etc.), if you know the relative importance of each characteristic, and if you can measure a number of potential solutions against this 'ideal' profile, then in principle you can <u>calculate</u> which potential solution fits best.

WAREHOUSE SPACE

FLOOR AREA — 3000 SQ F

HEIGHT — 21 FT

ACCESS M1 — 3.5 MILES

COST — £2.75 PER SQ FT PE

LEASE — 5 YR RENEWABLE

This can be very useful, but you must be clear about its limits — the calculation only applies to the comparison. The original judgement of which characteristics matter, and how important they are, is still wholly personal — a wonderful solution for me may be a monstrosity for you. Nor is there any guarantee that valued characteristics will be measurable — is there really a simple measure of 'customer satisfaction', 'pleasant environment', 'personal enjoyment', 'loving relationship' . . .?

Even if it works, there is a very big difference between recognising the logic of a choice, and gaining active organisational commitment to it. Organisations are political, not logical, entities.

Contraction by personal choice.

If you are lucky, you may be free to choose for yourself.

Then the progression over several cycles of expansion and contraction is the story of your personal search and exploration.

Just on market, located only a short walk from shops buses at Temple Fortune, an attractive semi-detached spaci property with south facing garden and 2 dble beds and 1 sing bath, sep, w.c., large lounge with attractive inglenook firepl and beamed ceiling, dining room, large kitchen, guest cl Garage and 100ft south facing garden. Recommended by

But creative thinking is easily destroyed...

consumer requir
main contingenc
parameters of
ongoing situa
solution t

... and can itself
be destructive.

71

So introducing new ideas can be a very uncomfortable business . . .

... you've got an idea ...

. . . you may be feeling a bit vulnerable.

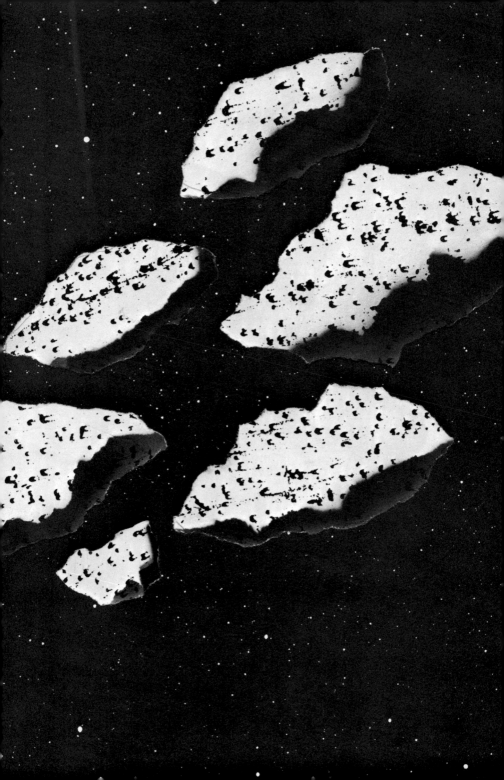

Finding the niche . . .

An idea with no substance behind it remains a pleasant, but ineffective, day dream.

An idea gathers substance when it begins to move into the centre of other people's attention, begins to affect their actions, and begins to have to fit in with other things around it.

There will always be lots of other projects claiming to be just as urgent, or valuable, or significant, for people to put their energy and money into, so finding and securing that niche is certainly not easy or automatic. You are going to have to explain, persuade, educate, negotiate, twist arms, befriend, disentangle misunderstandings, defend, attack . . . You may even find it easier to compromise, and join with someone with a related idea who already has a 'place of their own'.

In practical terms, your project will need a number of things:

—a financial and technological 'power supply'.

—ripe conditions and timely action

—the confident support of the decision makers

—the personal enthusiasm of the participants.

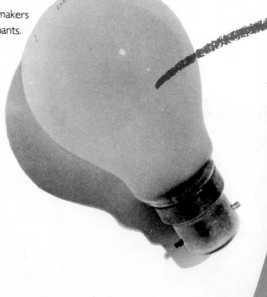

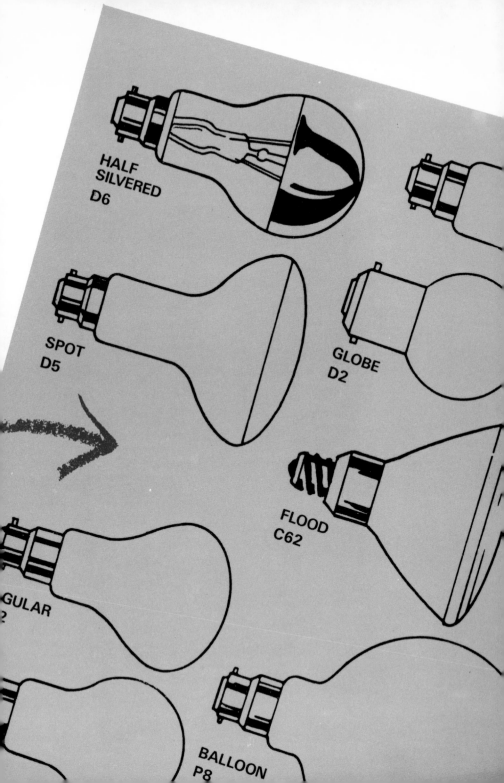

HALF
SILVERED
D6

SPOT
D5

GLOBE
D2

GULAR

FLOOD
C62

BALLOON
P8

Tapping the power supply.

Sometimes new ideas will just 'catch on', spreading like rumour, because they meet widespread needs or interests.

But more often, you will need the backing of an organisation to 'turn on' your project into a living, productive, process.

Power is the ability to determine which projects the organisational machine will back — which ideas will be passed over, and which will have resources put at their disposal.

The power to channel resources may be exercised by individuals, because of their strength, wealth, status, or expertise. Or it may come from the collective reactions of groups of people, expressed through democratic processes, popular movements, or market forces.

The resources that power releases often have concealed problems that you did not expect. Money may have to be spent in one way and not others, and by a certain time; the available skills may not be those you want; there may be additional agreements, regulations and obligations of all sorts.

So though you need a power supply and the resources it brings, choose it carefully — the wrong voltage or frequency may blow the bulb.

And there are other ways you have to fit in as well ...

Ripeness

Ideas 'take off' most easily when they give a shape to something that is already 'trying to get out'. 'Ideas' people are like artists — both try to express possibilities that are just under the surface in the world around them.

If the technical, organisational, social and personal conditions are not 'ripe' even the best idea will need a lot of 'power' to get anywhere, and the systems in which it is embedded may be severely distorted in the process. Change is much easier and less destructive when people are ready for it. If you want to row a fragile boat across a fast river you must find ways to let the current help you rather than trying to battle against it.

Flight has been around as an idea for thousands of years. But only this century have the necessary technology and resources and the needs for long distance travel come together. The vision of a democratically responsive world is also an old one. Is the time ripe?

If changes work best when conditions are ripe, they will often be small. But they don't have to be. The caterpillar, the chrysalis, the butterfly and the egg, are about as different as they could be — but each emerges from its predecessor when the conditions are ripe. If the tensions are large, they may resolve easily into a very different form. But deeper structures are often harder to change quickly, except in chaotic ways. A butterfly won't change into a moth.

Of course if you want a more radical change than the system is ready for, there is nothing to stop you trying to speed up the process of ripening!

COMING
SOON:
PLATO
SWIFT
LOCKE
PAINE
MARX
GANDHI
FANON

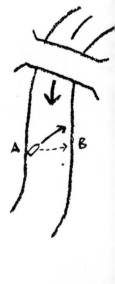

complex system is a complex machine.

It has all sorts of internal cycles, and sequences — committee cles, accounting periods, academic years, projects that start and ish at certain times, contract deadlines, sequences of operations at mustn't be interrupted once started, personal career cycles of gnificant members, periods of optimism or gloom, and so on.

If you want to adjust this 'machine', you have to match your ovements and timing to it.

If you don't, you are likely to miss the bus.

If you catch the organisers at the right points in a system of ommittee cycles, you may be able to get through a whole series of olitical stages in a fortnight. A couple of days later, and it may take onths, during which all sorts of things may change.

You need to have your finger on the many pulses of the rganisation, and know how they inter-connect. Often there are nly a handful of people who have the necessary understanding of e machinery and they will not necessarily be willing to put their ill at your disposal.

Trust and uncertainty

Change is never completely predictable.

When an organisation is mobilised behind a particular project, nobody knows for sure what will happen. So the decision-makers have to make a lot of guesses.

As well as gambling calculated risks against possible payoffs, they will be under the influence of many conflicting and changing pressures. Much will depend on their judgements about your judgement — and their ability to sway other people's judgements about their judgements about your judgements! A lot of your ground-work involves establishing your own credentials as someone worth listening to.

After all, why <u>should</u> anyone take <u>your</u> advice? Why should anyone even take time to listen to you? (or us!)

If people <u>do</u> agree to go along with you they are taking a lot on trust.

So they need to know a lot about you, and you need to know a lot about them. But knowing what someone <u>can do</u>, is not the same as sensing what they <u>are</u> . . .

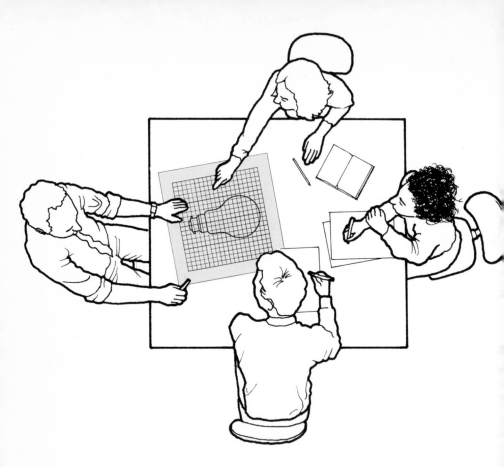

Doing

If you are going to chair a committee, the members want to be sure you have the organising and political skills to do it, so that you don't waste their time. If you are designing a product, you have to have the technical skills to be sure that it can be manufactured and will work. If you are preparing a contract, the parties involved will want to know that you are skilled in contract law.

These are skills to do with the task at hand. They are usually learned skills — activities you have become good at. One can judge this sort of skill by looking at samples of someone's work, or by talking to people who have made use of it, or by evidence of relevant training or experience or qualifications.

Systems people tend to know a little about each of many topics — enough to understand how all the different specialist areas in any one project fit together, but not enough to do the specialist jobs themselves. Part of the ground-work is to identify what skills will be needed, and to assemble a team that has them.

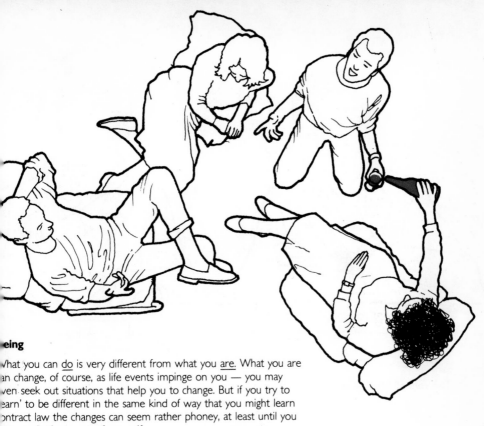

eing

What you can <u>do</u> is very different from what you <u>are</u>. What you are
an change, of course, as life events impinge on you — you may
ven seek out situations that help you to change. But if you try to
earn' to be different in the same kind of way that you might learn
ontract law the changes can seem rather phoney, at least until you
ave made them part of yourself.

People show what they <u>are</u> by how others react to them.
pending time with people who are at peace with themselves and
alue listening to you feels very different and much safer than being
ith people who are very tense, and seem driven by inner demons
at are quite unaware of your existence.

A team of people trying to put an idea into practice face many
nxieties, some of which are objective ("will X be delivered on
me?"), some personal ("can I really do it well enough?"). There will
ften be periods when no-one can distinguish between realistic
ars, and unfounded panics.

If the atmosphere in the team feels safe and trustworthy, the
nfounded panics and anxieties can easily be checked out and
osorbed. But if external pressures such as job insecurity, or internal
ressures from 'inner demons' are too great, the anxieties may easily
ke over, with people becoming suspicious of one another,
efensive manoeuvering, scape-goating, leadership conflicts,
stracting side-issues, and all the paraphernalia of a group at war
ith itself.

Ground-work includes maintaining a good atmosphere, with good
ommunication and trust. You may have to start by changing
ourself.

Non-committal scepticism . .

Sullen anger . .

The role of 'change agent'.

As an agent of change you must expect such responses. It is in the nature of the role, and the change agent needs the skills, adaptability, resilience, and groundedness, to cope with it.

Some of the pressures and uncertainties within an organisation are transmitted in from the turbulent world outside, and some are generated from inside by the tensions between the needs of the organisation as a whole, and the needs of the individual members within it. But wherever the pressure comes from, every sectional interest expects to have to defend itself, to 'guard its patch' from real and imagined threats, and change is almost always a potential threat.

There is no single 'correct' way to be a change agent. Many different styles are possible. There are 'high profile' styles that 'lead from the front' such as the entrepreneur, the visionary, the autocrat and 'low profile' ones that 'work from the back' such as the non-directive facilitator, the under-cover agent, the adviser. Some aim to sell particular ideas; others aim to help people to find their own ideas. Some aim to provide information: others aim mainly to inject energy and vitality; others to increase the level of trust and communication.

It will certainly help if you can let the angry confront you; fire enthusiasm in the bored and disillusioned; provide a safe place for the hurt, the tense, and the anxious to open out in; hear what the arrogant have to say, without being pushed around by them; satisfy the sceptical; and explain carefully and respectfully to the confused.

Withdrawn boredom . .

Distracted disinterest . .

MODESTY

P AG NCE

DIPLOMACY

So far it has all been preparation. But eventually something more tangible has to happen . . .

5. implementation

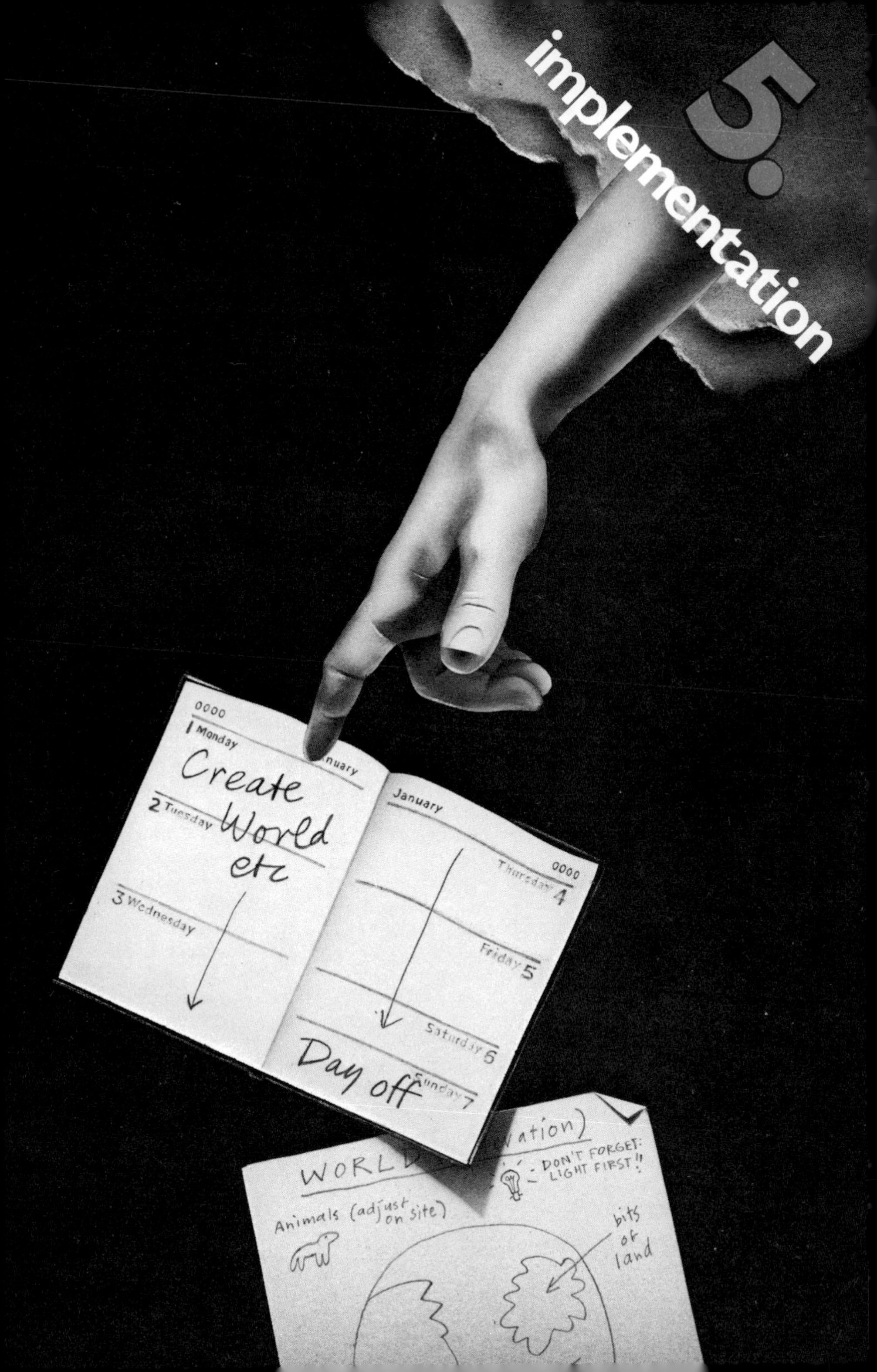

Buying a motorbike is rather lik

You are going to buy a bike! Great! Just before you sign the cheque, do you mind answering a few simple questions — it won't take a moment . . .

What bike do you want? What size should it be? How fast? How expensive? What special clothes will you need? Have you got a crash helmet? What about a driving licence? And have you remembered the insurance? And don't forget the log book. And the road tax licence of course. And where will you keep it? Have you got a padlock? Would you like some tools? And how about a repair manual? And will you need a fuel can? How will you carry your brief-case and your sandwich box? And what about your umbrella and handbag? And of course you will need to have access to a helpful garage or workshop. Will you pay cash? Or instalments? Will you want a loan? Do you want to buy the accessories now? Or perhaps later, to spread the cost? You don't want a big outlay just before the holidays, do you!? Can you afford it at all? But will you be able to use it without accessories? But the way, have you considered the guarantee period? And what about third party liability? And could the shop repossess it? And where are you going to keep the mower if the bike is to go in the garden shed? It will be a fire-risk in the hall, and do you mind the smell of petrol? Of course, then there is work — what will the boss think of you when you walk in like a spaceman? Mind you the others might find th more human than a two-piece suit. Have you got permission to the car park? Do you think it might be risky for the kids to r and do you want the older ones to get ideas abo ᵗ ʰⁱ fit are you — you can get ᵧ ᵤ cold on a hik accident — how's your ᵗ friend᷈

onducting a symphony orchestra

Like to hear Wagner's 'Meistersinger' would you? Great! Just a few things to get together before we start — it won't take a moment . . .

One piccolo, two flutes, two oboes, two clarinets, two bassoons, four horns, three trumpets, three trombones, bass tuba, two kettle drums, or triangle, cymbals, one harp, two banks of violins, one h bank of cellos, one b~~~~ ~f d~~~i~~~ ~~~ ~~~ ~~~ house, two .

Every project needs both 'specialist' and 'ensemble' skills and everyone in the project team needs both to some extent, though you often find a split between those who concentrate most on being specialists, like the orchestral players, and those who concentrate most on the ensemble, like the conductor. You need both.

Every project has its own specialisms. We can't tell you in this book what they might be in your case, not because they aren't important (quite the reverse — they are vital) but because you can't tell us what your project is.

But we can tell you a little about the ensemble. Here are some of the systemic ensemble skills that are often needed . . . and some of the problems . . .

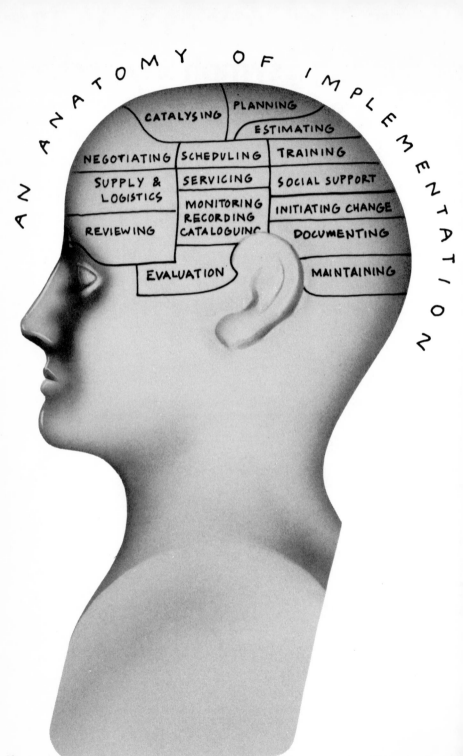

AN ANATOMY OF IMPLEMENTATION

CATALYSING
PLANNING
ESTIMATING
NEGOTIATING
SCHEDULING
TRAINING
SUPPLY & LOGISTICS
SERVICING
SOCIAL SUPPORT
MONITORING RECORDING CATALOGUING
INITIATING CHANGE
REVIEWING
DOCUMENTING
EVALUATION
MAINTAINING

The questions . . .

Catalysing: How can we get people going, energise, develop interest, build the team?

Planning: How can we break the work into manageable chunks that fit together successfully and minimise long-term cost?

Estimating: What resources will we need, in what quantities, at what cost?

Scheduling: When, and in what order, should we do things to minimise crises and wasted effort?

Supplying and logistics: How do we get the particular resources needed where they should be, when they should be; get rid of what is not needed; and avoid putting things that are not yet needed where they will get in the way?

Negotiating: How do we agree with the people involved what they should do for us, and what we should do for them?

Training: How can we make sure that people can do what we want them to do?

Servicing: What about the practical human needs of those involved? — meals, toilets, first aid, overalls, relaxation, travel . . .

Social support: And what about the personal needs of those involved? — emotional, social, developmental . . .

Monitoring, recording and documenting: How do we keep track of what tasks have been completed, financial records, stock levels, outstanding problems, agreed decisions? What about legal records, quality control records, accident reports, training manuals, fault-finding guides, contracts, terms of reference, certificates, licences?

Reviewing: Stepping back to see the overall state of play:— are we meeting targets, are problems emerging, do we need to redirect our resources?

Maintaining: Once things have been set up, how can we keep them going? — machines, buildings, organisations, procedures — particularly when they have been in operation for a long time, and the initial enthusiasm and alertness has worn off.

Evaluation: Stepping back even further. Is the system really doing what we thought it would do? Is it still the best way of doing it? Do we still need it at all?

Initiating change: Identifying and carrying out necessary adjustments; envisaging and initiating new uses of resources; closing one system down in order to form a very different one.

The problems . . .

MURPHY'S LAWS:
Reality never fits, and it's your fault.

1ST LAW:
IF A SYSTEM CAN'T FAIL— IT WILL

2ND LAW:
EVERYTHING TAKES LONGER THAN YOU THINK

3RD LAW:
IF IT LOOKS O.K.— IT ISN'T

4TH LAW:
IF THERE ARE TWO SHIFTS, THE OTHER SHIFT DID IT..

93

First of all, our special thanks to . . .

. . . everyone who takes the trouble to replace the loo rolls, check holiday dates, answer the telephone, find the elastoplast, notice the leak, water the plants, remember meetings, look after visitors, paint the building, tidy up, buy pencils, keep the notice-board up to date, arrange the dental appointments, return the jacket left behind, buy tickets for Vancouver, provide an ear for angry people to blast off at, adjust the furniture, check the supply of disposable gloves, get the blocked pipe freed, answer an unexpected phone call, look after people who fall ill, put in a new light bulb, find some ash-trays, put away the cups, close the windows, pick the car up from the garage, remind people, find lost addresses . . .

There is no such thing as a foolproof routine. Reality is always much, much more complex than even the best blueprint; hence the wry 'Murphy's Laws'. Not even the most brilliant of systemic designs can provide a routine substitute for alert, interested, adaptability.

Nevertheless, a well-designed routine is enormously better than a bad one . . .

I am not going to type any more of this for you until you learn to write NEATLY !!!

Building a co-ordinated system

The elements of control for any task are, of course:
—Knowing what is supposed to happen
—Monitoring its progress
—Comparing its actual progress with its planned progress
—Doing something about it if these differ

If a project is small enough for one person to have a clear overview of what is happening, control systems need not be elaborate — a conductor can control an orchestra, a works manager can control a workshop . . .

But in large projects, co-ordination between sub-systems and over time, and maintenance within sub-systems, become crucial problems,

How should a major project best be divided up to areas of responsibility, or phases of work?

How can the 'ensemble' controllers have an overall picture of where each sub-task has got to, its progress towards its goal, its use of its allocated resources?

How can one identify problems as they arise in such a complex system, and what can one do about them, particularly if they fall between different areas or phases?

How can each sub-system maintain its integrity within the overall system?

It is useful to distinguish four types of control:

Longitudinal co-ordination over time.

Cross-sectional co-ordination at a particular point in time.

Detailed logistics and scheduling.

Defending and maintaining structural boundaries.

city

Olympic stadium 'will be too late'

Montreal, Nov 5.—The Montreal French-language newspaper *La Presse* said today that a Quebec Government committee had concluded it would be almost impossible for the main Olympic stadium to be completed in time for the opening of the 1976 games here on July 17.

La Presse reported that members of the committee, headed by Mr Fernand Lalonde, the Quebec Solicitor-General, had

met the games organizing committee on Monday and recommended that the organizers should begin looking for an alternative site for the track and field and swimming events. The swimming pool is part of the main stadium.

Officials of the organizing committee declined to comment on the *La Presse* report, but they said that Mr Roger Rousseau, president of the organizing committee, would be

discussing the position with the officials concerned.

Construction of the $375m main stadium has been delayed periodically over the past 18 months by strikes and other labour disputes.

The last walk-out, which ended two weeks ago, caused a complete stoppage of work for about a week.

The latest cost projection is $730m and this is regarded by many experts as still too low.—Reuter.

Girl dies after

Stadium 'in time for Olympics'

Montreal, Nov 6.—The stadium for the 1976 Ol next July will be re but may lack " trimmings "

Mr Quebec S

Lalonde, Governm g committ t quite exa eport in M hat his com it would be to have the completed in ng of the Ga

the Quebec night that the thought it prud Games organizi to define its stages for the Games so completely finished, we could still hold the Games.

The latest cost estimate of $730m is still regarded as too low by many experts. The organizing committee has it expects a deficit $250m.—R

ntains that Lower House it a year. Sir Governor-pressure to dy in a situ-dence offers

The Whole-life

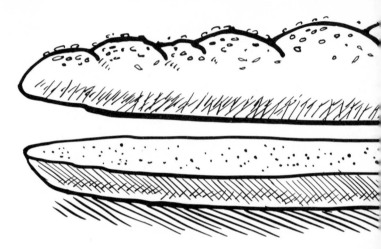

Longitudinal co-ordination over time

The idea here is to optimise financially and operationally to ensure the lowest <u>average</u> expenditure, hassle, etc. over the <u>whole life</u> of the project from conception to obsolescence and replacement, even though this may mean a bit of extra effort or expenditure at some points.

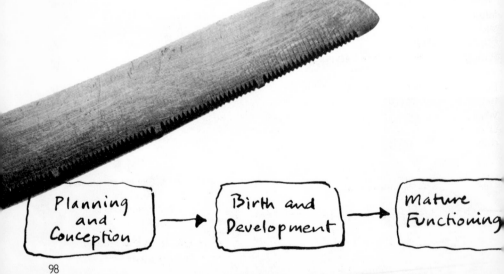

Planning and Conception → Birth and Development → Mature Functioning

Viewpoint.

A basic checklist for longitudinal ensemble:

Progress monitoring: what stage are we at now? What stages are supposed to be complete already, and how complete are they? What later stages are we expecting, and are the rudiments of any of them visible yet?

Transitions: what is needed to smooth the transition from the previous stage into this one, and from this stage into the next?

Information: what information has been handed on from previous stages? What information should you be handing on to later stages, how, and to whom?

Problems: what problems have we inherited? What problems will we be passing on?

Room for manoeuvre: are you passing on a resilient enough project to permit some redirection if circumstances change?

Whole-life costing: to what extent can future costs be reduced by extra costs now? To what extent will current savings result in extra costs later?

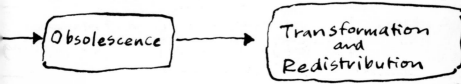

Obsolescence → Transformation and Redistribution

A Slice of Life.

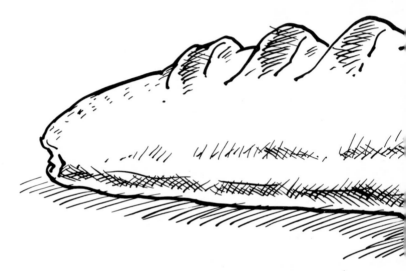

Cross-sectional co-ordination at one point in time

Large projects have to be divided up. But this fragmentation creates many local boundaries and interfaces, which form a common area for problems of non-compatibility or poor communication to arise in. Components made in one sub-system may fail to plug in successfully to sockets in equipment made by a different sub-system; messages that are issued by one sub-system may be distorted, lost, or misunderstood on the way to a second sub-system, and so on.

The problem here is to design and maintain an organisational structure in which each sub-system gets on efficiently with its own specialism, but yet the whole is co-ordinated into a coherent ensemble.

If the operations involved were routine and highly structured, then you would start by carefully dividing the overall project into component tasks, providing exact specification rules for each; each sub-system would then carry out its own component task according to those rules; eventually, if the rules were correct, the results of each task would fit together correctly to complete the project.

However the greater the uncertainty about how to break the overall task into well-defined components, the higher the proportion of effort that will go into continuing discussion and co-ordination throughout implementation, rather than just during the initial planning.

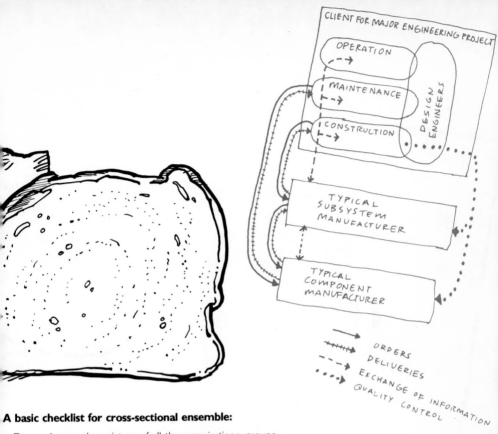

CLIENT FOR MAJOR ENGINEERING PROJECT

OPERATION

MAINTENANCE

CONSTRUCTION

DESIGN ENGINEERS

TYPICAL SUBSYSTEM MANUFACTURER

TYPICAL COMPONENT MANUFACTURER

ORDERS
DELIVERIES
EXCHANGE OF INFORMATION
QUALITY CONTROL

A basic checklist for cross-sectional ensemble:

Do you have a clear picture of all the organisations, groups, facilities, etc that make up your implementation system, the sub-tasks they have been allocated and the environments that affect them? Is this structural division satisfactory?

Do you and they know how they are supposed to connect together? What level of connection is appropriate between each pair of sub-systems? Are some inter-connections more critical than others?

How is each connection monitored and checked, and do the connections work out as they are supposed to do?

If structural divisions or inter-connections don't work, what can you do to improve them?

May

23
Monday 143–222 Week 21

✳ Set up marquee

✳ check guest list

24
Tuesday 144–221 Week 21

Install beer in marquee and allow to settle

25
Wednesday 145–220 Week 21 FAYE week 8

Wine and cheese arrives today

Thursday 146–219 Week 21
○ Full Moon

GARDEN
PARTY

2pm
Sta

Friday 147–218 Week 21

10am: Sandwiches delivered

Saturday 148–217 Week
sr 4.53, ss 21.03

Sunday 149–216 W
Trinity Sund

102

So what am I supposed to do with 3,000 cucumber sandwiches?

Logistics and scheduling

Think of a large construction site, with lots of sub-tasks, each of which depends on earlier sub-tasks; later sub-tasks depend, in turn, on them. Space is limited, so you must neither store material on a spot that is about to be built on, nor store materials that are to be used soon underneath those that won't be needed for ages. Co-ordinating and sequencing all this to minimise unnecessary delays and expense is no easy job; and changing weather conditions, late deliveries, site problems, etc., mean that the schedule has to be revised continually.

Another checklist:

Do you know what the sub-tasks are, and how they depend on one another?

What is the 'critical path', or sequence of stages that have no option but to be done one after another, and which therefore limits the overall rate of progress? These must be the top priority activities.

Where are the actual and potential bottle-necks?

What scope have you for re-jigging your time-table to cope with the inevitable surprises?

How do you communicate the time-table to everyone concerned?

How do you find out whether the time-table is going as planned?

How do you deal with unreliable information? It is only human for people to say they can do more than they achieve in fact, or to present their failures in a favourable way.

Defending and maintaining boundaries

You can only co-ordinate a set of sub-systems if the structure of each sub-system remains clear enough so that you know what you are co-ordinating. If the local structure within sub-systems breaks down, then an overall structure between sub-systems will be meaningless.

The local structure is maintained partly by scaled down longitudinal, cross-sectional, and logistic control, applying within the sub-system rather than within the system as a whole.

But another part is the maintenance of boundaries, and their defence against encroachment or degeneration, often in a highly competitive and invasive setting. Organisations can easily lose their alertness and responsiveness, and be trampled over by accident, vandalised, taken over for other purposes, blocked, swamped, subverted . . .

Subtle approaches to defence may use quite limited power and
resources since they use the energy of the attacker, not of the
defender — delaying, querying, disrupting, subverting, sabotaging.

But 'high power' approaches, such as actively destroying or
blocking, require the defender to have the power and energy to
confront the attacker.

The difficulty with defensive activity is in knowing when it is really
necessary, and when it is the defence itself that is creating the
enemies it was supposed to be aimed against. This is less likely if you
can maintain communication with the attackers at the same time as
defending against them. Once communication breaks down, you are
on your own with your imagination.

Team building

But planning, co-ordinating, scheduling and the defence of boundaries are not enough on their own.

They all depend on being able to persuade people to work together, to co-operate, and to be committed to the project.

Synergy is the state in which the team 'takes off', working together as a whole to achieve far more than the individuals, working separately, could have done.

Its opposite has been called emotional plague — the unpleasant state when everyone seems to be wilfully at cross-purposes with everyone else, and the group achieves much less than the individuals, working separately, could have done.

Team building can use every human skill you have. It needs understanding, informing, educating, choosing, inspiring, inducing, negotiating, arm-twisting, caring, confronting, seducing, leading, listening, following, reflecting, containing . . .

TEAM

**A true-life epic
of challenge,
commitment and
collaboration**

A final checklist:

Whose help do you need?

What do they need?

What will raise their enthusiasm?

What will bring out their skills?

What prevents them from listening to one another,
and building on one another's support?

Why should they work with you?

Books

General books with excellent bibliographies:

Systems thinking, systems practice. Peter Checkland, Wiley, 1981 ISBN 0 471 27911 0.

The Revised All-New Universal Traveller: A soft-systems guide to creativity, problem-solving and the process of reaching goals. Don Koberg and Jim Bagnall, Kaufman, 1981. ISBN 0 86576 017 9.

Peter Checkland is the Professor of Systems at Lancaster University, and is the leading academic exponent of applied soft systems work in the U.K. This is a scholarly and thorough book, the first half of which reviews the history and philosophical roots of the soft systems approach while the second half reports on the Lancaster departments' very extensive practical experience with Peter Checkland's own methods.

The Californians Don Koberg and Jim Bagnall adopt a very different approach, which was one of the inspirations for our book. They have provided a hand-book of problem-solving techniques set in a soft systems framework, with a very lively and well documented style of presentation.

Some other books you might find interesting

Systems behaviour (3rd edition), Open Systems Group (eds.) Harper & Row, 1982 ISBN 0 06 318212 2

Organisations as systems. Martin Lockett & Roger Spear (eds.) Open University Press, 1980. ISBN 0 335 00263 3

Organisations: cases, issues, concepts. Rob Paton et al. (eds.) Harper & Row, 1984 ISBN 0 06 318264 5

Intervention theory & method: a behavioural science view. Chris Argyris, Addison-Wesley, 1971. ISBN 0 333 26838 5

Thinking in organisations. Colin Eden et al. Macmillan, 1979

Kalila and Dimna — Selected fables of Bidpai. Ramsay Wood, Paladin (Granada), 1982. ISBN 0 586 084096.

The Vickers Papers. Open Systems Group (eds.) Harper & Row, 1984. ISBN 0 06 3182270 X

The Art of Judgment. Sir Geoffrey Vickers, Chapman and Hall, 1965, re-issued by Harper and Row, 1983.

The first three of these are sets of readings that have been prepared for Open University Systems courses (see below). Argyris' book is a classic in the field. Colin Eden's book describes an approach being developed within the Bath University School of Management that we feel very sympathetic towards, though Colin Eden himself would be reluctant to have his work labelled 'Systems'. The two books by the late Sir Geoffrey Vickers present another 'soft systems' approach compatible with that described in this book. 'Kalila and Dimna' is included for the reasons mentioned on page 24

Courses

University courses in Systems are available in the U.K. at Aston University School of Management (Birmingham), City University (London), Lancaster University, and the Open University. Most of the material for Open University courses is publicly available in libraries and bookshops and all of its Systems courses can be studied either as part of an undergraduate degree (enquires to: Admission Office, Open University, Milton Keynes, MK7 6AB) or as independent modules within the associate student programme (enquiries to: Associate Student Office, Open University, Milton Keynes, MK7 6AN). The relevant courses are:

Systems Behaviour (code No: T241) is a second level course consisting of a series of case-studies of different kinds of system, and teaching systemic methods for describing them.

Living with Organizations (code No: T244) is another second level course that looks at organisations, and activities within them, from a systemic point of view.

Complexity, management and change: applying a systems approach (code No: T301) is third level (honours) course that develops the material taught in T241 and T244 by developing systems methods for practical use in organisational development, decision making and decision analysis. This book was designed as an introduction to this course.

Glossary

Abstract system — A system whose components are theoretical or conceptual, e.g. the grammar of a language, a set of formal rules, or a mathematical theorem.

Action — Something that somebody does knowingly, consciously, and deliberately. For instance turning out the light is an action, but snoring when you are asleep is behaviour, not action. An observer who does not know what you are thinking sees your behaviour, but can only guess which parts of it are conscious actions, and which are unconscious habits.

Adaptive control — A closed loop control mechanism (q.v.) i.e. one involving feedback. The continuous or intermittent adjustment of some activity so that it matches changing conditions.

Agent (or Agency) — The person (or group of people) who initiates, carries out or mediates action (q.v.).

Analysis — A phase of the agent's activity which concentrates on trying to understand a situation.

Awareness — The first stage in the Method for System Description, at which you become aware of some situation as being of interest or concern to you, and can perhaps give some vague description of it, but cannot yet articulate clearly the nature of your concern, or the detailed nature of the situation.

Boundary — The notional demarcation line used by an analyst or observer to separate a system from its environment. It acts as a rule which indicates whether some entity is inside the system or not.

Client — The person (or group of people) who commissions the agent and with whom terms of reference are negotiated. In some cases this may involve a formal contract with fees; in other cases the agent may simply be instructed to do the work.

Climate — A mental and emotional framework of attitudes, expectations and habits shared by participants in a human activity system (q.v.) and experienced by them as a background influence to many of their actions. It is often profoundly affected by how the functional needs of the system's process (q.v.) fit within its structure (q.v.). .

Closed loop control — A control mechanism where part of the output of a system is returned to the input in such a way as to affect the input or some of the operating characteristics of the system. This is also called adaptive control (q.v.).

Closed system — A system which has no interaction with its environment. In their 'pure' form such systems exist only as theoretical concepts. There are, however, relatively closed real systems, in which significant interaction with the environment is very low.

Commitment — The second stage in the Method for Systems Description, which results in a statement of your reasons for your interest in a particular situation, that is clear enough to guide the general direction of your analysis, and to measure your progress against. It will not give you detailed terms of reference, but it will give you a clear answer to "why am I doing this?".

Comparison — A key step in any control process is to compare the adjustment of the activity being controlled to some standard criterion level or model. In non-adaptive control (q.v.) this only happens once, when the activity is set up. In adaptive control (q.v.) it happens intermittently or continuously, and includes information about the recent performance of the activity.

Component — A recognisable part of a system which may be a sub-system or an element (q.v.).

Controls — The mechanisms which act to preserve the relationship between the structure and process of a system. In other words, they tend to maintain the system in existence as a recognisable entity under changing circumstances.

Designed system — A system deliberately planned and constructed by one or more people, usually to achieve a particular set of objectives, e.g. a car assembly line.

Detection — Some versions of the Method for System Description (q.v.) use this instead of Testing (q.v.).

Digraph convention — In mathematical terms, a digraph is a network of lines joining pairs of points. The basic element in a 'digraph' is therefore:

There is a body of mathematical theory about digraphs, but digraphs that are simple enough to be readily understood by a general audience (as is usually the case in practical organisational analysis) do not usually warrant the more sophisticated analysis that the mathematical theory offers.

Element — A system component which, at the current level of analysis, you do not intend to sub-divide any further.

Environment — This is the set of elements that affect the system, but are not controlled by it. Though clearly relevant to the system, they are regarded as falling outside its boundary (q.v.).

Feedback — A characteristic of 'closed loop control' (q.v.). The modification of a variable, process, or system in the light of its own effects or outputs. In the strict sense, the modification depends on the difference between the actual state and the reference state, but it may be used more loosely to refer to any causal loop.

Feed-forward — A characteristic of 'open-loop control' (q.v.). Called feed-forward because it is essentially control based on prediction of the future rather than reaction to the past.

Gate-keeper — A person (or group of people) who controls access to channels of communication or to particular resources or procedures.

Goal — Target for medium to long-term strategies. Usually in general rather than specific terms. A direction to move in, rather than a detailed quantitative objective (q.v.).

Groundwork — A phase of activity which concentrates on ensuring that a proposed activity is acceptable to those involved in it, and will be actively backed by them.

Hard system — A system which is mainly designed and quantifiable, with well defined objectives. Mechanical and technical systems are often of this type, but highly routinised human systems can sometimes be treated in this way as well.

Heuristic approach — Tackling a problem on the basis of learning by experience, trial and error, and intuition.

Hierarchy — An inverted tree structure, in which the upper levels have fewer elements, but each is of greater status, or scope, and the lower levels have many elements, of lesser status or scope. For instance, a bureaucratic hierarchy, a hierarchy of levels of analysis, or a hierarchy of objectives.

Human activity system — A system in which the main components (q.v.) are people and their actions (q.v.). Because they are bound to involve subjective perceptions, these systems are almost always best treated as soft systems (q.v.).

Implementation — A phase of the agent's activity which concentrates on the actual execution of changes that have important consequences and cannot easily be reversed.

HOT COLD

Inventive search — A phase of the agent's activity aimed at exploring, developing, and selecting possible actions in ways that are flexible and reversible so that trial-and-error is an acceptable approach.

Iteration — The repeated application of a set of procedures, or stages of a method, to the same material, with the intention that each pass will produce some further progress towards a full solution.

Method for system description — A method for developing an understanding of a complex situation which involves stages of: awareness (q.v.), commitment (q.v.), testing (q.v.), separation (q.v.) and selection (q.v.) of systems, followed by detailed analysis and description. The stages may be iterative (q.v.) and may precede the design and implementation (q.v.) of change.

Model — A description of a real or hypothetical situation, usually formal and simplified, which is used to develop understanding.

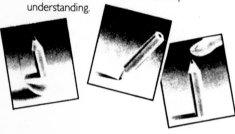

Natural system — A system whose components are naturally occurring, e.g. the blood circulation system, or the behavioural and physiological system that controls breeding in a particular species.

Non-adaptive control — See 'open loop control'.

Objective — Short-term, practical and specific target. The tactics (q.v.) for achieving it may be closely prescribed. Much more detailed than a goal (q.v.).

Objective description — An attempt at rational, dispassionate description of a situation, independent of the values and beliefs of the analyst or observer.

Open loop control — A control mechanism where performance is pre-set and there is no automatic corrective action (apart from possible external intervention). This is also called 'feed-forward', 'predictive', or 'non-adaptive' control (q.v.).

Open system — A system which interacts with an environment, i.e. where there is an exchange of energy, information, or materials across the boundary.

Optimising — System objectives often conflict, in the sense that if one is achieved it will prevent another one being achieved. A central part of decision-making is 'optimising' — i.e. deciding which compromise between conflicting goals will give the greatest benefit on balance. 'Satisficing' is a weaker form, in which the decision is merely to choose a compromise that can be accepted, and that meets certain minimum requirements.

Policy — A very broad statement of a particular value system. A statement of ideological beliefs, ethical principles, or terms of reference. Long term goals (q.v.) may be derived from statements of policy.

Problem-owner — The person (or group of people) in whose sphere of action, or area of responsibility, the problem situation is focussed.

Process — Change. The events, actions communications, and changing relationships in a situation.

Purposeful control — The capacity of a system to devise alternative courses of action, to display free choice amongst them, and then to pursue the selected action in adaptively controlled ways. Purpose with choice.

Purposive control — The capacity of a system to pursue a pre-set goal in an adaptively controlled way. Purpose without choice.

Relationship — A is said to have a relationship to B when A's behaviour is affected by the presence or absence of B.

Resolution — Because of limitations in our capacity to process information, analyses cannot be *both* detailed *and* broad. Any analysis must be conducted at an appropriate compromise between breadth and depth. This is its level of resolution.

Rich picture — An evolving diagram which collects together and portrays key information and impressions about a complex situation, in a loosely structured and evocative way. This is also called a situation summary.

Ripeness — This is the notion that a system will vary in its readiness to change in a particular way not only because of short-term cycles of activity, but because of longer-term developmental processes. At each stage in its development, certain types of change are 'near the surface' and can be implemented easily, whereas other types of change can only be imposed with great difficulty, and may impose considerable stresses on the system.

Selection — The fifth stage of the Method for Systems Description, in which the analyst selects and names one or two key relevant systems from the broad array of systems identified in the situation which the analyst is working on. They are selected as the areas most likely to repay the use of further (limited) analytic resources. Selection and separation (q.v.) are usually very closely coupled, with frequent iteration (q.v.) between them.

Self-maintaining — The ability of a system to hold itself in existence (usually in a relatively stable state) over a long period of time (unless it is destabilised by radical change).

Separation — The fourth stage of the Method for Systems Description, in which the analyst collects information about the chosen situation, and identifies a number of areas of it which can be described in systems terms, which seem relevant to the task set at the commitment stage (q.v.).

Serendipity — The art of making happy and unexpected discoveries by accident.

Situation summary — See 'Rich Picture'.

Soft system — A system depending largely on non-routinised human actions (q.v.), so that human capacity for free choice, and the agent's (q.v.) limited access to the subjective values, beliefs and wishes of the participants means that wholly objective description (q.v.) or quantitative modelling are not appropriate.

Strategy — A medium or long-term programme of activities related to particular goals (q.v.). A strategic programme usually leaves flexibility for later adjustments and choice of detailed tactics (q.v.), but it will probably set broad directions, outer limits, and general guidelines.

Structure — The setting and relationships in a situation that can be regarded as more or less 'fixed' within the time-scales the analyst is concerned with.

Subjective description — A description of a situation which includes and is affected by the values and beliefs of the analyst or the person making the description.

Sub-system — A system component above the chosen limits of resolution (q.v.) so that it may contain within it sub-sub-systems, or elements (q.v.). A smaller unit within a system which could also be viewed as a system in its own right, at a finer level of resolution (q.v.).

Synergy — The effect produced when the operation of a whole appears to be greater than the sum of the operation of each of its parts.

System — A recognisable whole which consists of a set of inter-dependent parts. More specifically:

a) A system is an assembly of components, connected together in an organised way.

b) The components are affected by being in the system and the behaviour of the system is changed if they leave it.

c) This organised assembly of components does something.

d) This assembly as a whole has been identified by someone who is interested in it (e.g. the agent, the client, or the problem owner (q.v.)).

Systematic — Carried out in a planned and orderly fashion. Not to be confused with 'systemic' (q.v.).

Systemic — Using systems ideas; treating things as systems or from a systems viewpoint; pertaining to, a system. Not to be confused with 'systematic' (q.v.).

Tactic — A short-term local plan of specific activities in connection with a particular project. Usually clear-cut and prescriptive, with very limited scope for flexible redirection.

Testing — The third stage of the Method for Systems Description, in which the task to which the analyst is committed (q.v.) is examined to see if a systemic analysis is necessary or appropriate.

Venn convention — Venn diagrams (named after the mathematician who invented them) are a simple way of representing the logical relationships of inclusion, exclusion, overlap, and membership:

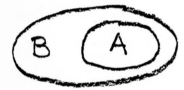

means: A is part of B (and there is a part of B that is not A). For instance my foot is part of me but there is a lot of me that is not my foot.

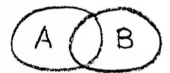

means: that there is a part of A that is also part of B. For instance there is a part of my life that is also a part of my wife's life but both of us have parts of our lives that do not overlap.

means: A and B are completely separate. For instance the Borough Councils of Milton Keynes and Aberdeen don't have much to do with one another.

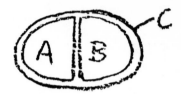

means: C consists of A and B and nothing else, and no part of A is also part of B. For instance my car can be thought of as consisting of the fixed parts and moving parts, and nothing else.

Because of the two-dimensional limitations of a printed page, complex overlapping of more than three entities can be impossible to represent in this way, and has to be handled by using the mathematics-like notation called symbolic logic. In Systems, we are dealing with very complex entities for which a strict adherence to Venn conventions may not work, and we have to use a certain amount of artistic licence!

Weltanschauung — The particular image of the world, outlook or world-view which colours and characterizes the perceptions of a person (or group of people) regarding any situation.